SpringerBriefs in Applied Sciences and Technology

Manufacturing and Surface Engineering

Series editor

Joao Paulo Davim, Aveiro, Portugal

More information about this series at http://www.springer.com/series/10623

Anouar Hajjaji · Mosbah Amlouk
Mounir Gaidi · Brahim Bessais
My Ali El Khakani

Chromium Doped TiO$_2$ Sputtered Thin Films

Synthesis, Physical Investigations and Applications

 Springer

Anouar Hajjaji
Research and Technology Center of Energy
Hammam-Lif
Tunisia

Brahim Bessais
Research and Technology Center of Energy
Hammam-Lif
Tunisia

Mosbah Amlouk
Faculty of Sciences of Bizerte
Bizerte
Tunisia

My Ali El Khakani
National Institute of Scientific Research
Varennes, QC
Canada

Mounir Gaidi
University of Sharjah
Sharjah
UAE

ISSN 2191-530X ISSN 2191-5318 (electronic)
SpringerBriefs in Applied Sciences and Technology
ISBN 978-3-319-13352-2 ISBN 978-3-319-13353-9 (eBook)
DOI 10.1007/978-3-319-13353-9

Library of Congress Control Number: 2014955615

Springer Cham Heidelberg New York Dordrecht London

Printed on acid-free paper

Springer International Publishing AG Switzerland is part of Springer Science+Business Media (www.springer.com)

Acknowledgments

This material is based upon work supported by the collaboration between two institutions: «Laboratoire de Photovoltaïque, Centre de Recherches et des Technologies de l'Energie, Technopole de Borj-Cédria, Hammam-Lif, Tunisia» and the Institut National de la Recherche Scientifique, INRS-Énergie, Matériaux et Télécommunications, Varennes, Canada.

The authors would like to thank all the people who kindly helped them, especially Dr. Samir Saidi for his help in the dissertation, Prof. Latifa Bousselmi for photocatalysis experiments, and Dr. Wissem Dimassi for his help during LBIC measurements.

Contents

Introduction

Due to its interesting intrinsic properties, Titanium oxide which belongs to the metal transition oxide family was the most studied during the last two decades and demanded material in many fields of applications such as transparent electrodes, gas sensors, heterojunction solar cells, photocatalytic process, etc. To improve the performance of this oxide, doping TiO_2 with suitable dopants offers an effective method to adjust some of its physical properties. Generally, the doping of semiconductors with appropriate metals is one of the most effective ways in research for developing sensitivity applications. However, the interaction between the doping metals and the semiconductor is complicated because the interaction relates to the carrier concentration, defect level, and surface states of the semiconductor, electronic, optical properties, and so on. Therefore, good understanding of the interaction will facilitate the fundamental and technical application of such oxides doped with some metallic elements.

It is indeed reported that this oxide based on transition metal and doped with an appropriate metallic element (Al, Nb, Sn, Ge, Fe, Cr) has a significant role in sensitivity applications such as photovoltaic solar cells, photocatalysis, and pollution sensors.

To achieve high efficiency in photovoltaic solar conversion, for example, solar cells-based TiO_2 as transparent conductor oxide (TCO) must absorb a maximum amount of light energy.

On the other hand, in the field of gas sensors, the characteristic response of a polycrystalline semiconductor device (for example the TiO_2) can be modified and controlled by several factors as: the size of particles, pore structure, density of grain. Knowledge of the influence of these parameters on the final component is essential for its operating.

Previous studies show that the variation of the microstructure is always accompanied by a more or less important variation of the electrical properties of the studied material. Metal additions are also an important factor; these additions can change dramatically the nature of the response of the material. However, it is important to distinguish the effect of these additions to the effect of other factors (microstructure, humidity…). In fact, to perform a reliable study of the effect of metal additions, it is important to set the microstructural parameters, and to have a series of samples prepared under the same conditions.

The first chapter deals with some basic principles on the operation and the performance of the components used in gas sensors, photocatalysis, and photovoltaic cells. We will focus on the effect of the microstructure and the incorporation of the metallic aggregates on the optoelectronic and sensing properties of Cr-doped TiO_2 films.

In the second chapter we give a brief introduction to the deposition technique of TiO_2 thin films by sputtering (used at the INRS-EMT labs, Canada). We will study the optoelectronic properties of Cr-doped TiO_2 thin films obtained by sputtering for use in the field of pollution and photovoltaic sensors.

In the third chapter, different analytical and microstructural characterizations of TiO_2 thin films prepared by sputtering method have been provided. We study those prepared in ambient temperature and also those annealed under oxygen at different temperatures.

In the first part of the fourth chapter, the electrical properties of the Cr-doped TiO_2 thin films have been studied. The measurement of the electrical conductance under ethanol vapors were carried out in terms of Cr content, in low and high concentration doping. In the second part, we studied the microstructural and optoelectronic properties of TiO_2 thin films deposited on the monocrystalline and multicrystalline porous silicon.

Abstract

Titanium oxide TiO$_2$ thin films have attracted tremendous research zeal in recent years. Sensitivity devices using this oxide depend on the structural, optical as well as electrical properties of such films. First, some of the physical properties, fabrication processes as well as the physical applications of TiO$_2$ material have been presented wherever necessary from previous works reported in the literature. Second, the synthesis protocol based on sputtering process has been detailed. A noticeable change from anatase to rutile phases of this oxide has been verified by means of some physical investigations such as: XRD, XPS, FTIR, reflectivity, ellipsometry, LBIC. It is also shown that the increase of doping content (from 0 % to 17 %) decreases the band gap energy value from 3.31 eV to 1.89 eV. Cr content controls the electrical conduction sensitivity of TiO$_2$ films under ethanol test. A change of type (N type to P type) is indeed observed for the two Cr concentrations 13 % and 17 % atomic. This behavior may be of interest for various sensitivity applications. Finally, concluding remarks are provided to bring out some clues regarding the possible use of such materials in some optoelectronic devices (photocatalytic, gas sensors, passivation of solar cells.).

Keywords Cr-doped TiO$_2$ · Thin films · Crystallographic structure · Microstructure · Band gap energy · Light Beam Induced Current (LBIC) measurements · Reflectivity · Photoluminescence · Photo-conversion · Gas sensors · Photocatalysis

Chapter 1
TiO$_2$ Properties and Deposition Techniques

Abstract This chapter deals with some physical properties of TiO$_2$ used in various physical applications. Indeed, a brief review of the structural, optical and electronic properties of TiO$_2$ films is presented; then some technical methods as well as fundaments and experimental features of this oxide are provided. Particular attention is paid to the effect of the microstructure and the incorporation of doping elements on the optoelectronic and sensing properties of TiO$_2$ films.

Keywords TiO$_2$ · Thin films · Crystallographic structure · Physical and chemical methods · Band energy diagram

1.1 Introduction

In this chapter, we present the context in which our work is developed supported by many bibliographic examples. In the first part, we detailed some of the crystallographic, optical and electronic properties of TiO$_2$ films. Then, we present the main methods for the synthesis of TiO$_2$ films. Finally, we give a number of basic principles on the operation and the performance of the components used in gas sensors, photocatalysis, and photovoltaic cells in the last part. We will in particular focus on the effect of the microstructure and the incorporation of the metal aggregates on the optoelectronic and sensing properties of TiO$_2$ films.

1.2 Crystallographic, Electronics and Optics Structures of TiO$_2$

1.2.1 Crystallographic Properties of TiO$_2$ [1]

Titanium dioxide exists in various forms including the anatase, rutile and the brookite. The rutile form is a thermodynamically stable phase at high temperature.

A. Hajjaji et al., *Chromium Doped TiO$_2$ Sputtered Thin Films*,
SpringerBriefs in Manufacturing and Surface Engineering,
DOI 10.1007/978-3-319-13353-9_1

Fig. 1.1 Phase transition of the titanium oxide

(a)   **(b)**   **(c)**

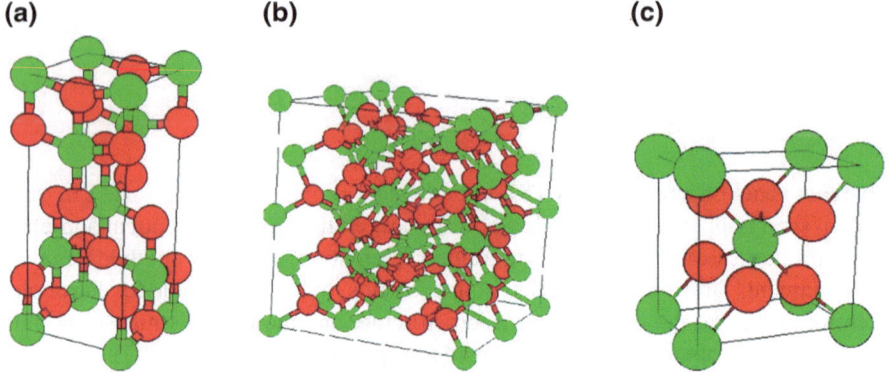

Fig. 1.2 Crystallographic structures of TiO$_2$ (Ti in *green*, O$_2$ in *red*)

Table 1.1 Crystallographic data of anatase, rutile and brookite TiO$_2$

	Anatase	Rutile	Brookite
Structure	Tetragonal	Tetragonal	Orthorhombic
Space group	$I\frac{4_1}{m}md$	$P\frac{4_2}{m}nm$	*Pbca*
Lattice parameter (Å)		a = 4.5930	a = 5.4558
		c = 2.9590	b = 9.1819
			c = 5.1429
Z	4	2	8

Below are represented the temperatures representing different titanium oxide phase transitions, Fig. 1.1.

The different crystallographic structures of the TiO$_2$ material are represented in Fig. 1.2 and Table 1.1 summarizes their major structural characteristics.

Only anatase and rutile phases are of technological interest. These two phases crystallize in the tetragonal system. In both structures, the titanium atom is surrounded by six oxygen atoms and each oxygen atom is surrounded by three atoms of titanium.

1.2.2 Electronic Properties of TiO$_2$

Titanium dioxide is a native n-type semiconductor having a wide bandgap of 3.2 eV [2] for anatase form and 3.0 eV [3] for rutile. In this semiconductor material, a light energy absorption hυ ≥ Eg (Eg TiO$_2$ = 3.2 eV, which corresponds to a wavelength below 380 nm) generates electron-hole pairs. The conductivity of the

Fig. 1.3 Diagram of the
energy bands

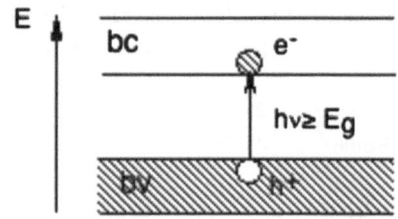

semiconductor depends on several intrinsic and extrinsic factors. Among intrinsic factors, one note the doping by 'donor' or 'acceptor' impurities of electrons; the energy levels of these impurities are located in the bandgap of the solid, near the limits of the conduction band (donors) and valence band (acceptors). If the thermal conditions are satisfied a donor may release an electron (e^-), and reach the conduction band. Similarly, an acceptor is able to capture an electron from the valence band and left a hole (h^+) considered as positive charged mobile carries (Fig. 1.3).

TiO$_2$ is a solid with predominant ionic character, consisting of Ti^{4+} and O^{2-} ions:

$$Ti\ 3d^2\ 4s^2 \rightarrow Ti^{4+}\ 3d^0\ (electronic\ structure\ of\ Ar) \tag{1.1}$$

$$O\ 2p^4 \rightarrow O^{2-}\ 2p^6\ (electronic\ structure\ of\ Ne) \tag{1.2}$$

The overlap of 2p levels, fully populated with oxygen ions, leads in the oxide to the formation of the valence band, while the overlap of 3d orbital of titanium ions leads to the formation of conduction band of the solid. From the microscopic point of view, the intrinsic absorption of light by this solid corresponds to a transition 2p (O) $\rightarrow$ 3d (Ti) [4].

1.2.3 Optical Properties of TiO$_2$

Titanium dioxide belongs to the chemical family of oxides of transition metals. It is produced industrially from the 20th century as a pigment for white paints, to replace the highly toxic lead oxides. It has a particularly high refractive index (n = 2.70 at λ = 590 nm for rutile). His insensitivity to visible light, due to its wide band gap (3.2 eV for the anatase) allows it to absorb in the near-ultraviolet. It presents a very high diffusion coefficient in visible light without absorption area (Fig. 1.4) [5].

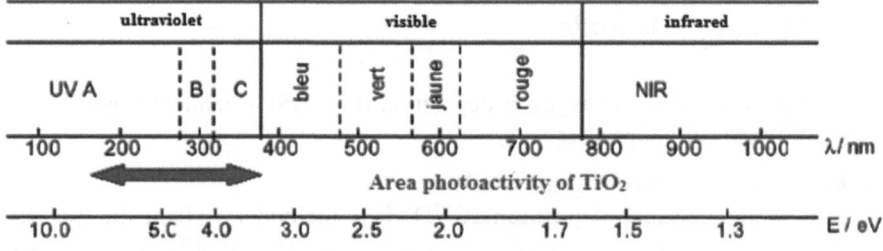

Fig. 1.4   The light spectrum with the zone of action of TiO$_2$

Table 1.2 Crystallographic data of anatase, rutile and brookite TiO$_2$

Phase	Refractive index in the visible	Density (g cm^{-3})	Crystallographic structure
Anatase	2.49	3.84	Tetragonal
Rutile	2.903	4.26	Tetragonal

Pure TiO$_2$ oxide is active by electromagnetic radiation of energy greater than Eg = 3.2 eV, corresponding to a wavelength less than 387 nm, located in the ultraviolet. However, UV radiation represents only 5 % of solar radiation. A simple idea to move the sensitivity of the TiO$_2$ to the visible area, and thereby increasing its photoactivity, is to reduce the bandgap and/or introduce discrete energy levels.

The optical properties of titanium dioxide depend on the technique of preparation of which depend on its refractive index, and the size of the grains. Its wide bandgap leads to strong absorption in the ultraviolet, that gives excellent protection properties against UV A and UV B. TiO$_2$ is the main active component of currently used sunscreens. TiO$_2$ is used, due to its very high optical index and its transparency in the visible and near IR, in most of the optical coatings anti-reflective layers for ophthalmic lenses or interference filters for optical telecommunication applications. Table 1.2 gives a comparison of the optical properties of TiO$_2$ films for both crystallographic forms: anatase and rutile.

1.3 Implementation Technique of the TiO$_2$

Several techniques of growth such as chemical methods, the sputtering and evaporation have been used for the deposition of TiO$_2$ thin films. However, the stabilization of its structure is sensitive to the deposition conditions such as the temperature of the substrate or the partial pressure of the used gas. Works on the fabrication and characterization of TiO$_2$ are very numerous. This part presents a review of some of the techniques used in the fabrication of TiO$_2$ thin films.

Figure 1.5 shows the deposition methods of TiO$_2$ thin film, which are organized around two types of processes: chemical and physical.

1.3.1 Chemical Methods

In this process we find two types of deposition (Fig. 1.5), Chemical vapor deposition (CVD) and sol-gel process.

(a) **Chemical vapor deposition (CVD)**
   The chemical vapor deposition (CVD) is a method in which the gas particles react to form a solid film on a substrate. The volatile compounds of

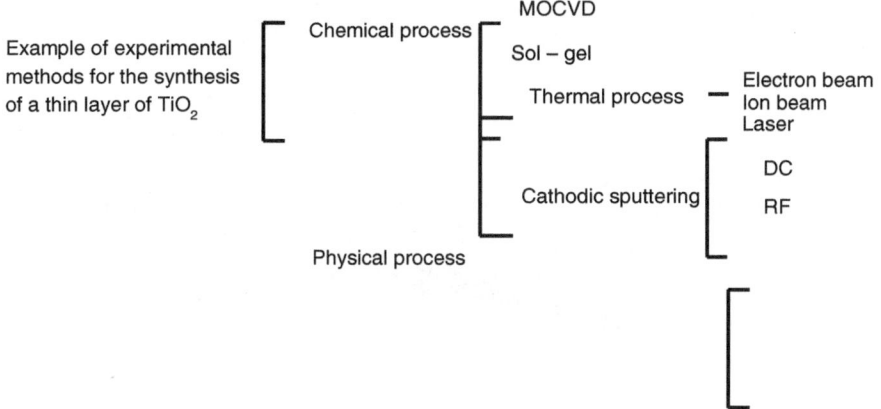

Fig. 1.5   General methods of deposition of TiO$_2$ thin films

the material to be deposited are diluted in a carrier gas and introduced in a reaction chamber where the substrates are placed. The film is obtained by chemical reaction between the vapor and the heated substrate. In some cases, increasing temperature is necessary to maintain the chemical reaction. The CVD is an interdisciplinary process; it includes a set of chemical reactions, thermodynamics and kinetics processes and transport phenomena [6]. The chemical reaction is at the center of these processes, it determines the nature and type of the present species. Two types of reactors exist: the hot wall reactor and the cold one. In the case of the hot wall reactor, this one is heated directly, allowing operating at lower pressure: at nearly 75 mTorr, in this case, thin-film deposition occurs on the substrates, but also on the walls (LPCVD technique: Low-Pressure Chemical Vapor Deposition [7]). In the cold wall reactor case, only the substrate is heated, so the reaction is effective at the level of the substrate and occurs in atmospheric pressure. The principle of this deposition method is presented in Fig. 1.6, in the case of hot-wall. We summarize the advantages and disadvantages of CVD method in Table 1.3.

(b) **Synthesis by Sol gel [8]**

This process is to synthesize an amorphous inorganic lattice by a chemical reaction in solution at room temperature. The used chemical precursors are often of alcoxydes organometallic-type, because they are very soluble in common solvents. Their hydrolysis rate is easily controllable and they are presented in an inorganic monomers form. The different steps of the sol-gel process are described in Fig. 1.7. The precursor is dissolved in a solvent under agitation. The solution is then hydrolyzed after contact with water. After hydrolysis, reactive monomers condense and form a sol. After aging and thickening of this sol, a gel is formed. It is then dried to remove traces of solvent and organic impurities.

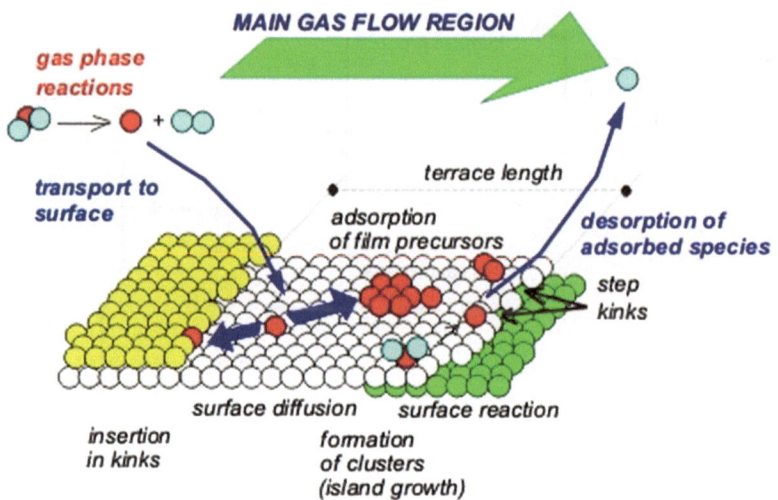

Fig. 1.6 Main steps of the chemical vapor deposition (CVD)

Table 1.3 Some advantages and disadvantages of CVD technique

Benefits	Disadvantages
Technical relatively easy to implement	The substrate must be heated
High deposition rate	Toxicity and aggressiveness precursors
Control of the stoichiometry, morphology and the crystal structure of the deposits	Contamination carbon (MOCVD)
Recovery supports uniform high dimensional complex shapes and hollow	High price of certain precursors of satisfactory purity
Power continuously without interruption of the vacuum in the reaction chamber	CVD is a technique of expensive deposit

Despite the significant advantages of this method, as the purity of the deposition and the good switching properties, we decided to not use it. Indeed, it sometimes gives inhomogeneous layer and not obtaining the desired thickness of nanometer order. In the next section, we present other fabrication process of TiO$_2$ following the physical methods.

1.3.2 Physical Method [Physical Vapor Deposition (PVD)]

This type of deposition process has many advantages compared to the chemical ones as the formation of dense films, better control of pollution and the ease of the process.

(a) **Synthesis by PLD (Pulsed Laser Deposition)** [9]:
    Figure 1.8 shows the diagram of a thermal deposition process using the strike of a target by pulsed laser (PLD).

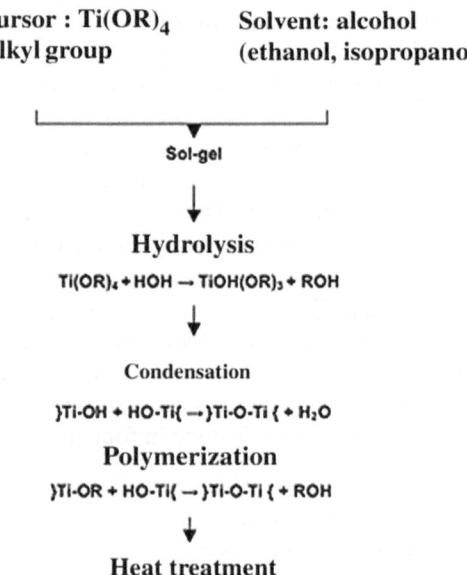

Fig. 1.7   Successive steps in a sol-gel synthesis

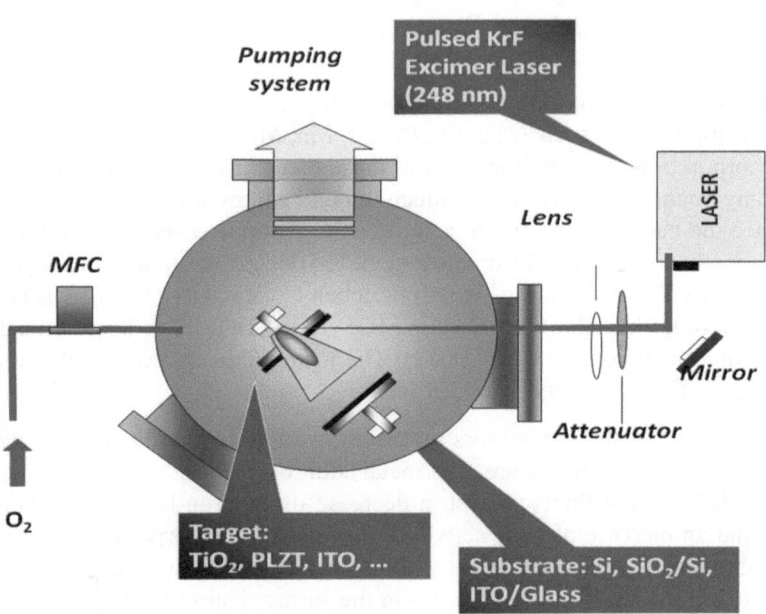

Fig. 1.8   Principle of laser ablation deposition (PLD)

The principle of laser ablation deposition is to evaporate material by focusing a laser beam of short duration and high power on the surface of a solid (Fig. 1.8). Over the ablation threshold, atoms, electrons and clusters are ejected from the surface and a very high density of particles and a high excitation temperature plasma appears. The laser fluence (energy per unit area) necessary to produce the plasma depends on the wavelength of the laser, the target material, and its morphology. The power can reach tens, or even hundreds of megawatts. Plasma subsequently condenses on a heated or not substrate.

(b) **Synthesis by sputtering [10]:**

The sputtering is a process in which atoms are extracted from the surface of a material due to a collision with particles at high energy. Inside a vacuum chamber a noble gas (argon) is introduced. Subsequently, under argon and oxygen atmosphere, the deposition is done in four steps:

1. Ions are generated and accelerated towards the target.
2. Generated ions bombard the target atoms.
3. Ejected atoms diffuse to the substrate.
4. Part of the sputtered atoms condense on the surface of the substrate to form a thin film.

This technique will be presented in detail in the Chap. 2.

1.4 Effect of Doping on the Optoelectronic and Microstructural Properties

One of the solutions to extend the absorption spectrum of TiO$_2$ involves doping with transition metals such as Fe, Cu, Co, Cr, Mn, Mo, Ni, Nb, Pt, W, UK [11, 12]. The incorporation of a small amount of these metals (0.1–5 % by weight) causes a rearrangement of valence and conduction band energy levels, leading ultimately to reduce the band gap. As result, a shift of the absorption spectrum to the visible domain is observed. Thus, Dvoranova et al. [13] demonstrate that the doped TiO$_2$ by chrome Cr(III), manganese (Mn(II) or cobalt Co(II) (0.2–1 % by mass) absorbs more in visible than the pure Degussa P25 TiO$_2$ (TiO$_2$ P25: 80 % anatase and 20 % rutile manufactured by Degussa). The metal addition leads to the improvement of sensitivity in various applications:

- Improvement of the gas sensors sensitivity with certain pollutants and decreasing the operation temperature after metal addition.
- In the field of the Photocatalyst, a decrease of electron-hole pairs recombination and an increase of TiO$_2$ activity in the near-UV. Incorporation of transition or noble metals decreases the recombination velocity by trapping e$^-$ and h$^+$. In addition, presence of metal particles in the surface can serve as anode and cathode increasing the charges transfers to the surface [14, 15].

1.5 TiO$_2$ Thin Films and Their Various Applications

Titanium oxide has many applications because of its chemical inertia, relatively no toxic quality, low cost, high refractive index and benefits related to its surface properties. It was introduced in the industry in the early 1900s, initially to replace the toxic white pigments. Indeed, it is used as a white pigment in decorative or architectural paints in wooden buildings, furniture and car industry. Its applications are much diversified. It is also used in photocatalysis, cosmetology, pharmacy, solar cells, waveguides, electrochromic systems, as a gas sensor, support or promoter of catalysis, etc.

1.5.1 In the Photovoltaic Field

TiO$_2$ is inexpensive, non-toxic, and relatively abundant and presents a high photoelectric response. It is used as antireflection layers in Silicon-based solar cells. Today, the TiO$_2$ is the best semiconductor oxide used in Gratzel-type solar cells [16, 17]. So far conversion efficiencies commonly exceed 3 % under solar irradiation, and reach 11 % for the most efficient system. Although the exact origin of TiO$_2$ efficiency compared to other semiconductor oxides is still poorly known, several factors have been advanced to understand this efficiency. One of these factors is the strong coupling between the surface and the metallo-organic or organic chromophores. This coupling is due to the "d" nature of the TiO$_2$ conduction band advantaging d–π* interactions. The high density of States in the conduction band also allows an entropy gain. High TiO$_2$ electrons effective mass (m* = Ame) disadvantages trapping of electrons in the intermediate states. This factor helps interactions with chromophore or oxidized mediators and thus limits the recombination phenomena.

Finally, TiO$_2$ has a high specific surface area allowing a large amount of dye adsorption. By comparison, SnO$_2$-based solar cells have a more moderate performance (on the order of 4.4 %). In this case, the effective mass of the electrons is weaker, which leads to the trapping of electrons in intermediate states and their recombination with oxidized mediator. For ZnO, the cell performances are twice lower than those reported for TiO$_2$. This difference could be explained by the fact that the rates of injection of the first excited state of the dye to the conduction band of ZnO are three orders of magnitude lower than those measured with systems based on TiO$_2$ [18].

In the Gratzel type cells, the absorption of light is provided by a single layer of dye chemically adsorbed at the surface of the semiconductor. After excitation by an absorbed photon, a well chosen dye, usually an organic complex [19], may transfer an electron to the semiconductor (injection). The electromagnetic field inside the material allows the extraction of this electron.

The positive charge is transferred to the mediator dye (redox) present in the solution which filled the cell (interception) and by this way transported to the counter electrode. By this last electron transfer, the mediator return to the reduced state and the loop is completed (Fig. 1.9). The TiO_2 anatase phase presenting a porous and nanocrystalline structure is the best photo-electrode in the dye cells (DSCs) because of its high specific surface area, which allows adsorption of a large amount of dye [20].

Controlling the size and the shape of the crystals, as well as the control of the optical gap energy of TiO_2 films by incorporation of transition metals (e.g. Cr) are the main parameters to optimize and to identify potential features of this material. One of the most interesting one to realize is to control the value of the bandgap to cover a wide range of the solar spectrum and improve the efficiency of the material.

1.5.2 In the Gas Sensors Field

Today, several varieties of pollutants are known for their harmful effects on the health and the environment. The contamination of the air by these pollutants is the consequence of human activity. These pollutants, in addition to the problems they can cause near their places of production, can migrate in the atmosphere and produce, by chemical reactions, secondary pollutants such as acid rain or ozone. Thus, regardless of the location, the air contains a variety of synthetic chemical products mainly emitted by industries and machines. The very strict safety regulations in domestic and industrial area, taken by the most industrialized countries, have opened channels for extensive research to design and fabrication of more

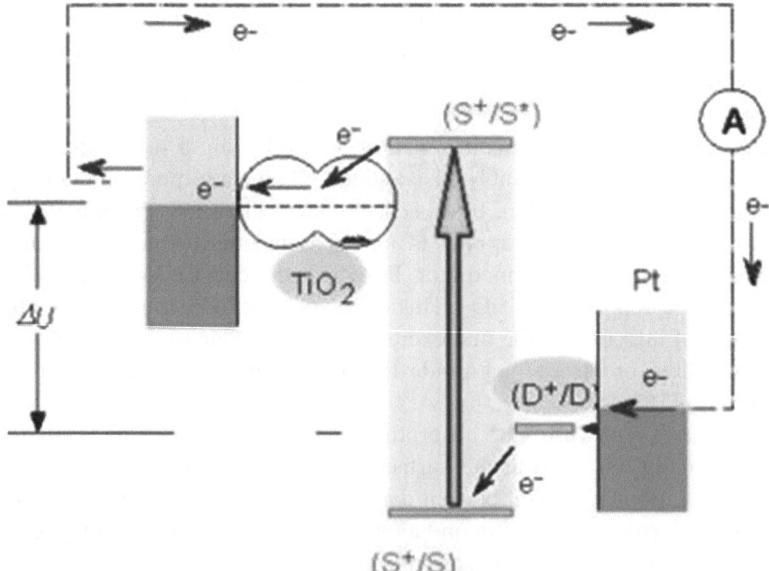

Fig. 1.9 Energy diagram of the solar cell state dye

efficient gas sensing systems. These sensors must be very selective, sensitive and reliable to be able to control various air pollutants. To achieve this goal, a wide variety of transducers materials have been exploited as: polymers, semiconductors [21–23] and many metal oxide semiconductors such as SnO_2, TiO_2, ZnO, etc.

Since then, SnO_2 type sensors have found wide rang of application in the field of pollutants gas sensing such as: NO, alcohols, etc. The principle of operation is always based on the reversible variation of electric conductivity during the adsorption of gases on the sensor surface. Despite of very important studies on Cr doped TiO_2 thin films sensors, the effect of metal additions remained so far poorly known. This constituted a serious obstacle for the development of sensitive and reliable sensors. A well understanding requires a detailed microstructural, analytical and optical analysis with respect to these films.

Chrome doping of TiO_2 layers has the following advantages:

- The response of the sensor in the presence of a pollutant gas is reversible and stable [24].
- Increased sensitivity [25].
- Stabilization of microstructure during the post deposition annealing.
- Possibilities of the conductivity control owing to Cr content [25].

1.5.3 In the Photocatalysis Field

Under the influence of Fujishima and Honda works [26] in the 1970s, on the photo-electrochemical behavior of metal oxides with large bandgap and the works of Frank and Bard [27] on the decomposition of cyanide in water, the photocatalytic technology has experienced a great boom. Since then, the Photocatalyst became an important studied domain, with the aim of increasing the environment protection, particularly water and air.

Titanium oxide has some interesting photocatalytic properties [28]. Indeed, when TiO_2 film is illuminated by photons of energy equal or greater than the band gap ($h\upsilon \geq Eg$), there is absorption of photons and creation in the solid of electron-hole pairs which dissociate into free electrons in the conduction band and holes in the valence band. These electrons can be used directly to create electricity in a photovoltaic solar cell [29], or cause chemical reactions known as the Photocatalyst [30]. Super hydrophilicity is another particular phenomenon that was discovered recently. It consists on trapping the holes on the titanium oxide surface causing a super wettability and giving to this oxide the name of smart or self-cleaning surfaces [31].

The semiconductors-based photocatalysis process, used for purification of air and water, has been developed around the titanium dioxide through the significant benefits that present this material [32]:

- It is stable, inexpensive, non-toxic;
- It is the most efficient Photocatalyst;
- It promotes the photodegradation of a wide range of pollutants at ambient temperature;

- The use of additives is not necessary;
- Inert from chemical and biological point of view.

1.5.4 TiO$_2$ Photocatalytic Operating Mode

The photocatalytic process is based on the excitation of TiO$_2$ film by light radiation of wavelength lower than 387 nm, which corresponds to energy greater or equal to the bandgap (3.2 eV):

$$TiO_2 + h\nu \rightarrow h^+ + e^- \tag{1.3}$$

An electron from the valence band is exited to the conduction band with the formation of a hole (h$^+$). The holes react with water and organic pollutants adsorbed on the TiO$_2$ film surface, following reactions (1.4) and (1.5):

$$H_2O + h^+ (valence\ band) \rightarrow OH^{\cdot} + H^+ \tag{1.4}$$

$$h^+ (valence\ band) + polluant \rightarrow polluant^+ \tag{1.5}$$

Hydroxyl radicals formed in reaction (1.5) also participate in the degradation of pollutants:

$$OH^{\cdot} + polluant \rightarrow CO_2 + H_2O \tag{1.6}$$

We must also consider electron-hole recombination reaction in TiO$_2$ surface or volume:

$$h^+ + e^- \rightarrow heat \tag{1.7}$$

The electron-hole pairs formation rate under photons impact (Eq. 1.3) depends on the intensity of the incident light and the optical and physical properties of the Photocatalyst. The charges diffusion rate to the crystallites surface is decisive for the formation of the OH$^{\cdot}$ radicals and therefore the degradation rate of the pollutant. The diffusion rate of the pairs and their recombination rate depend on several structural factors: The allotropic composition [33], crystallinity [34], crystallites size [35], and doping rate of ion [36]. These factors have indeed an influence on the photocatalytic activity. On the other hand, the photocatalytic activity depends on the pollutant chemical nature and the chemical complexity of its molecule.

Impurities in TiO$_2$ oxide film can play a beneficial role in the photocatalytic activity when their concentration is controlled. The effect of impurities integration in the TiO$_2$ film was studied by Park et al. [37]. These authors tested the photocatalytic activity of TiO$_2$ thin films prepared by the sol-gel technique and doped by low valence cations as Fe^{3+}, Co^{2+}, Ni^{2+} and high valence ones as: MB^{5+}, Nb^{5+}, W^{6+}. The photocatalytic activity of TiO$_2$ for trichloroethylene vapors conversion is significantly higher when the catalyst is doped with MB, Nb and W instead of Fe, Co and Ni. This difference is assigned to the high crystallinity of the sample doped with high-valence cations [38]. In the present work, we are interested in the choice of the Cr as dopant; the incorporation of Cr in the TiO$_2$ film can play a beneficial role in the photocatalytic activity.

References

1. Pighini C (2006) Syntheses de nanocristaux de TiO_2 anatase a distribution de taille controlee. Influence de la taille descristallites sur le spectre Raman et etude des proprieties de surface. Thèse doctorat, Université de Bourgogn
2. Chatterjee S (2008) Titania-germanium nanocomposite as a photovoltaic material. Sol Energy 82:95
3. Cronemeyer DC (1952) Electrical and optical properties of rutile single crystals. Phys Rev 87:876
4. Florence BOSC (2004) Synthese et caracterisation des couches minces et de membranes photocatalytiques mésostructurees a base de TiO_2 anatase. Thèse de doctorat, Montpellier II
5. Černigoj U, Lavrenčič Štangar U, Trebše P, Rebernik Ribič P (2006) Comparison of different characteristics of TiO_2 films and their photocatalytic properties. Acta Chim Slov 53:29–35
6. Vossen JL (ed) (1978) Thin film processes. In: Academic press; traite de la pulvérisation en général avec une liste de références très complète jusqu'à 1977
7. Bouchier D (1985) Thèse de doctorat, Orsay
8. Lakshmi BB, Dorhout PK (1997) Sol-gel template synthesis of semiconductor nanostructures. J Chem Mater 9:857
9. Long H, Yang G, Chen A, Li Y, Lu P (2008) Growth and characteristics of laser deposited anatase and rutile TiO_2 films on Si substrates. Thin Solid Films 517:745
10. Zakrzewska K, Radecka M, Rekas M (1997) Effect of Nb, Cr, Sn additions on gas sensing properies of TiO_2 thin films. Thin Solid Films 310:161–166
11. Demeestere K, Dewulf J, Ohno T, Salgado PH, Van Langenhove H (2005) Visible light mediated photocatalytic degradation of gaseous trichloroethylene and dimethyl sulfide on modified titanium dioxide. Appl Catal B Environ 61:140–149
12. Lee K, Lee NH, Shin SH, Lee HG, Kim SJ (2006) Hydrothermal synthesis and photocatalytic characterizations of transition metals doped nano TiO_2 sols. Mater Sci Eng B 129:109
13. Dvoranová D, Brezová V, Mazúr M, Malati MA (2002) Investigations of metal-doped titanium dioxide photocatalysts. Appl Catal B Environ 37:91
14. Chen J, Ollis DF, Rulkens WH, Bruning H (1999) Kinetic processes of photocatalytic mineralization of alcohols on metallized titanium dioxide. Water Res 33:1173
15. Yang P, Lu C, Hua N, Du Y (2002) Titanium dioxide nanoparticles co-doped with Fe^{3+} and Eu^{3+} ions for photocatalysis. Mater Lett 57:794
16. Pr. Thierry Toupance (2008) Thèse doctorat, Université du Bordeaux
17. Natsuhara H, Matsumoto K, Yoshida N, Itoh T, Nonomura S, Fukawa M, Sato K (2006) TiO_2 thin films as protective material for transparent conducting oxides used in Si thin film solar cells. Sol Energy Mater Sol Cells 90:2867
18. Liu B-Q, Zhao X-P, Luo W (2008) The synergistic effect of two photosynthetic pigments in dye-sensitized mesoporous TiO_2 solar cells. Dyes Pigm 76:327
19. O'Regan B, Gratzel M (1991) A low-cost, high-efficiency solar cell based on dyesensitized colloidal TiO_2 films. Nature 335:737
20. Tang X, Qian J, Wang Z, Wang H, Feng Q, Liua G (2009) Comparison of low crystallinity TiO_2 film with nanocrystalline anatase film for dye-sensitized solar cells. J Colloid Interface Sci 330:386
21. Holec T, Chvojka T, Jelinek I, Jindřich J, Němec I, Pelant I, Valenta J, Dian J (2002) Determination of sensoric parameters of porous silicon in sensing of organic vapors. Mater Sci Eng C 19:251–254
22. Barillaro G, Nannini A, Pieri F (2003) APSFET: a new, porous silicon-based gas sensing device. Sens Actuators B Chem 93:263
23. Barillaro G, Diligenti A, Marola G, Strambini LM (2005) A silicon crystalline resistor with an adsorbing porous layer as gas sensor. Sens Actuators B Chem 105:278
24. Li Y et al (2002) Gas sensing properties of p-type semiconducting Cr-doped TiO_2 thin films. Sens Actuators B chem 83:160

25. Ruiz AM et al (2003) Cr-doped TiO$_2$ gas sensor for exhaust NO$_2$ monitoring. Sens Actuators B Chem 93:509
26. Fujishima A, Honda K (1972) Electrochemical photolysis of water at a semiconductor electrode. Nature 37:238
27. Frank SN, Bard AJ (1977) Heterogeneous photocatalytic oxidation of cyanide and sulfite in aqueous solutions at semiconductor powders. J Phys Chem 81:1484
28. Carp O, Huisman CL, Reller A (2004) Photoinduced reactivity of titanium dioxide. Prog Solid State Chem 32:33
29. Longo C, De Paoli M-A (2003) Dye-sensitized solar cells: a successful combination of materials. J Braz Chem Soc 14:889
30. Kontos AI, Kontos AG, Tsoukleris DS, Valchos GD, Falaras P (2007) Superhydrophilicity and photocatalytic property of nanocrystalline titania sol-gel films. Thin Solid Films 515:7370
31. Karuppuchamy S, Jeong JM (2005) Super-hydrophilic amorphous titanium dioxide thin film deposited by cathodic electrodeposition. Mater Chem Phys 93:251
32. Zhao J, Yang X (2003) Photocatalytic oxidation for indoor air purification: a literature review. Build Environ 38:645
33. Schindler K-M, Kunst M (1990) Charge-carrier dynamics in TiO$_2$ powders. J Phys Chem 94:8222
34. Maeda M, Watanabe T (2007) Effects of crystallinity and grain size on photocatalytic activity of titania films. Surf Coat Technol 201:9309
35. Almquist CB, Biswas P (2002) Role of synthesis method and particle size of nanostructured TiO$_2$ on its photoactivity. J Catal 212:145
36. Choi W, Termin A, Hoffmann MR (1994) The role of metal ion dopants in quantum-sized TiO$_2$: correlation between photoreactivity and charge carrier recombination dynamics. J Phys Chem 98:13669
37. Park SESE, Joo H, Kang JW (2004) Effect of impurities in TiO$_2$ thin films on trichloroethylene conversion. Sol Energy Mater Sol Cells 83:39
38. Sarantopoulos C (2007) Thèse doctorat, Institut national de polytechnique Toulouse

Chapter 2
Synthesis and Characterization of TiO₂–Cr Thin Films

Abstract Details of the important experimental conditions of the synthesis of Cr doped TiO₂ thin films by means of co-deposition by magnetron sputtering process are presented. Furthermore, a brief description of the structural, electrical and optical characterizations set-up is provided. These physical characterizations are based on X-ray diffraction, atomic force and scanning electronic microscopy, LBIC… techniques.

Keywords Chromium–Doped TiO₂ · Magnetron sputtering method · Chemical etching · Physical characterization setup

In this chapter, the experimental methods and the experimental condition of the synthesis and characterization of TiO₂–Cr thin films have been detailed. The used fabrication processes are the co-deposition by magnetron sputtering and the electrochemical attack technique. We also give a brief description the structural, electrical and optical characterizations setup.

2.1 Synthesis of Thin Films by Magnetron Sputtering

The sputtering is one of the most used techniques for the preparation of the semiconductor metal oxides thin films. This technique allows both qualitative and quantitative control of the thin films deposition. The deposition experimental setup is presented in Fig. 2.2, it consists of an argon vacuum chamber, equipped with a system target-magnetron door and heated stage. This apparatus is powered by a radio frequency generator. Under the action of electric field applied by the radiofrequency generator, argon atoms are ionized and accelerated inside the target electric field. The target has a negative voltage compared to the argon plasma. The pulverized particles tend to be electrically neutral and distributed in the chamber. A number of them are collected onto a substrate placed in front of the target, and on which thin film is grown.

A. Hajjaji et al., *Chromium Doped TiO₂ Sputtered Thin Films*,
SpringerBriefs in Manufacturing and Surface Engineering,
DOI 10.1007/978-3-319-13353-9_2

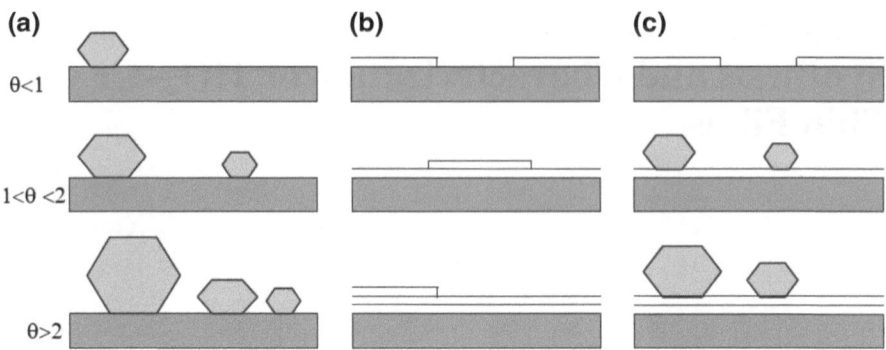

Fig. 2.1 The different modes of thin film growth

2.1.1 The Thin Film Formation Steps

2.1.1.1 Classification of the Growth Modes

In a simple approach, we class the thin films growth on a substrate into three categories [1] schematically illustrated in Fig. 2.1:

(a) **Patches growth (Volmer-Weber (VW) mode)**
In this growth mode, patches are formed on a substrate surface by clusters nucleation, Fig. 2.1a. This growth will take place when atoms or molecules, arriving on the substrate surface, have more affinity to bind together than to the substrate. Therefore, it is three-dimensional growth: a typical case of this growth is that of metal films on insulator substrates.

(b) **Layers growth (Franck-Van der Merwe (FM) mode)**
This growth mode takes place when the atom-substrate interaction is very strong. The first atoms arriving on the substrate surface condense and form a single layer covering the entire substrate area, Fig. 2.1b: there is a two-dimensional nuclei growth to form a layer, and so on layer by layer.

(c) **Mixed growth (Stranski-Krastanov (ST) mode)**
This growth mode is an intermediate case between the two previous modes (VW and FM modes): the growth is primarily two-dimensional to form the first thin film layer; However, as the energy of atom-substrate interaction gradually decreases, growth tends to become three-dimensional with patches formation, Fig. 2.1c.

2.1.1.2 The Morphology of Thin Films

The thin film morphology depends on different basic parameters as: the adsorption energy of deposited elements, their possible interaction with the substrate atoms, their mobility on the surface, sputtering rate, thermal diffusion, temperature, etc.

Fig. 2.2 System
co-deposition by magnetron
sputtering

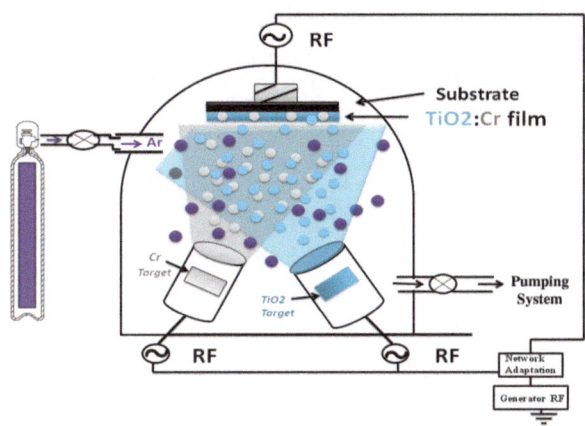

2.1.2 Principle of Operation

2.1.2.1 Description of the Deposition Machine

The used deposition chamber consists on a CMS-18 reactor including three sputtering posts. Co-disposition by magnetron sputtering system is schematized in Fig. 2.2.

Before the plasma switch on, the system must be pumped to reach a vacuum of order 10^{-5} mTorr. To produce the discharge, a partial pressure of high purity argon equals to 1.2 mTorr is created in the chamber. Targets can then be turned on individually, thus creating two independent plasma sources. The power of each RF sources is then adjusted within a range between 8 and 360 W depending on the desired composition.

2.1.2.2 Deposition Requirements for the Development of Cr-Doped TiO$_2$ Thin Films

Cr-doped TiO$_2$ thin films have been developed by cathodic radiofrequency magnetron sputtering of two targets: TiO$_2$ and chromium (Cr). Purity index of the two targets are 99.995 and 99.99 % for titanium dioxide and chromium, respectively. The applied power to the target, the pressure, flow and volumetric ratio of Ar and O$_2$ gases, as well as the possible polarization of the substrate are chosen based on the results of previous work (Table 2.1).

2.1.2.3 Drawbacks and Benefits

(a) **Benefits**

- Deposition of all types of materials: refractory metal (W, Ta, Mo, …), alloys, oxides, dielectrics.
- Good adhesion.

Table 2.1 Deposition conditions used for the preparation of thin films of TiO$_2$

Power applied to the TiO$_2$ target (W)	Power applied to the Cr target (W)	Ar flow (sccm)	O$_2$ flow (sccm)	Pressure (mTorr)
360	8	80 %	20 %	1.2
//	15	//	//	//
//	25	//	//	//
//	50	//	//	//
//	100	//	//	//
//	150	//	//	//

- Good uniformity of thickness if target diameter $\gg$ substrate diameter.
- This method allows getting large surfaces films.

(b) **Drawbacks**

- More expensive equipment and more difficult to implement compared to evaporation technique.
- The deposition is made under an inert atmosphere (Argon). The resulting film is not crystalline as in the vacuum deposition case.
- Defects are created in layers by energetic particles and plasma UV photons.

2.1.2.4 Conclusion

Magnetron sputtering is less efficient than the CVD process due to deposition velocity. However, its implementation is simpler and can remove any solid material at ordinary temperature. In addition, magnetron sputtering technique allows deposition of insulating layers as well as metallic layers (aluminum, chrome …). It is therefore widely used for the realization of interconnection layers in integrated devices.

2.2 Electroetching

For the development of thin layers of porous silicon (PS), we used the conventional technique of electrochemical anodization (AE). It takes place in a wet environment based hydrofluoric acid (HF). This technique consists of the silicon substrate to attack the anode of an electrochemical cell. During the attack, a fixed current flows between the anode (the starting substrate) and the cathode (gold or platinum).

2.2.1 Cell Anodizing

Thin films are produced in a cell comprising a single reservoir Teflon (Fig. 2.3). During the formation of SP, the rear face of the Si substrate acts as anode (it is connected to the positive pole of the power supply). Its front face (polished surface) is contacted with the electrolytic solution is maintained at a negative electrical

Fig. 2.3 Electroetching

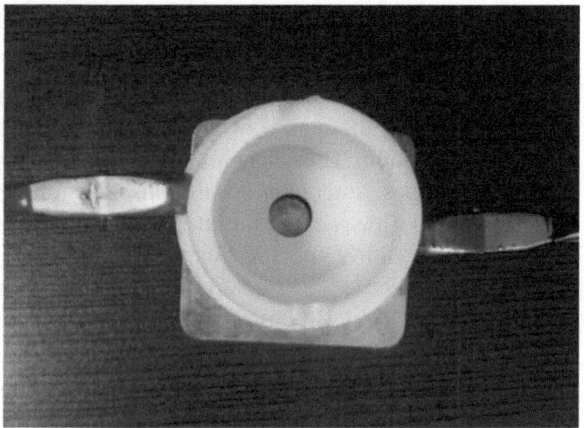

potential relative to the rear face. The rear contact is ensured by moderate pressure on a plate stainless aluminum.

The sealing of the silicon wafer is provided by an O-ring which delimits the area of formation of the PS (about 4 cm^2). The cell is connected to a potentiostat used as current source. A computer serves to control the variation of the anodization current.

2.2.2 Development of the PS Layers

Before the formation of the PS, the substrate is degreased. For this, the sample is rinsed successively in a solution of acetone and deionized water. These rinses help remove dust and organic origin of the surface of the substrate particles. To improve the homogeneity of the layers, an ohmic contact Si/Al on the back is made. The procedure consists of depositing an Al layer by evaporation on the rear face of the silicon wafer, followed by drying in an infrared oven under a nitrogen stream and baked for 15 min at a temperature of 550 °C [2]. The porous structures are obtained from a single silicon substrate or polycrystalline P-type, boron doped, orientation (100), respectively resistivity 1–3 Ω cm and 5–2 Ω cm having a thickness of order of 500 and 330 microns. The porous layers have a circular area of about 4 cm^2. The electrochemical etching is performed in hydrofluoric acid, HF (40 %) diluted in ethanol with 1:4 as ratio.

2.3 Microstructural and Analytical Characterization

In this section, we present the characterization techniques used to understand the microstructure and composition of thin-film-type TiO_2–Cr obtained by magnetron sputtering.

2.3.1 X-ray Photoelectron Spectroscopy (XPS)

X-ray photoelectron spectroscopy is a technique of surface analysis to obtain information about the nature of chemical bonds in the films as well as their elemental composition. Thus, when a thin film is exposed to X-ray photons, atoms that compose these interact with photons. This energy ratio leads to the expulsion of some of the electrons of the atoms constituting the heart layer, the electrons leave the atom with a specific kinetic energy. This, combined with the nature of the material concerned, can determine the distance traveled by the photoelectrons expelled by the X-ray. In case the average path can reach the surface of the film, they are extracted from the material and pass into the void where they are then collected by a system of magnetic lenses and/or to guide them to electrostatic spectrometer. They can then be counted and classified according to their kinetic energy, which is closely linked to their binding energy:

$$h\nu = E_L + E_C + W \tag{2.1}$$

Thus, since the photon energy hv X incidents and the work function W are known energy EL binding the electron can be deduced.

As part of this work, we used a VG 220i-XL Escalab device VG Instruments, the vacuum base is 10^{-10} Torr, combined with a hemispherical detector. Al kα a monochromatic source (1486.6 eV) and Mg kα (1253.6 eV) permit analysis of orbital Ti 2p, 2p Cr, O1s CasaXPS using the software and before and after cleaning the surface of the films by spraying by means of a beam of Ar$^+$ ions to 5 keV.

2.3.2 Infrared Fourier Transform Spectroscopy (FTIR)

This technique allows us to study the nature of chemical bonds in a material. For this purpose, the infrared frequencies (energy) absorbed by the chemical bonds of the molecules present in the sample are measured. These bonds have different modes of vibration corresponding to wavelengths specific infrared wavelengths, working in a wide frequency range to identify the types of bonds present in the film. In practice, if we work in transmission mode, the sample is exposed to an infrared transmitted beam which passes through an interferometer, a signal is then obtained as an interferogram. A Fourier transform of this signal is performed, resulting in a frequency spectrum. A first measurement is performed on a blank substrate (no deposit) and then the sample (thin film + substrate). For FTIR measurements, we used a Nicolet 6700 FT-IR unit of Thermo Electron Corporation, to vary the wavelength of far-infrared to the UV-Visible (400–4,000 cm^{-1}). The test chamber is constantly purged with ultra-dry air, which eliminates the absorption of water modes.

2.3.3 Grazing Incidence X-ray Diffraction (GIXRD)

The classical method of X-ray analysis typically involves an X-rays penetration depth from the micrometer to millimeter rang. Thus, measurements of X-ray diffraction (XRD) in the classic configuration θ–2θ are generally less adapted to the c of the deposited thin films crystalline structure (risk of interference with the substrate XRD diagram). For these reasons, measurements of X-ray diffraction are done with grazing incidence angle θ between the incident beam and the sample surface (on the order of 1°) [3]. Therefore, the irradiated depth decreases and the length of the X-ray path in the thin layer increases laterally, which allows to an increase in the intensity of the signal from the thin film. In this configuration, the incident angle θ remains fixed while the detector of the diffracted X-ray is mobile in 2θ (angle of deviation).

For our GIXRD measures, we used a PW3040/60 X pert Pro diffractometer from Philips equipped with a Cukα (1,543a) source, the incidence beam angle θ is set to 0, 5°. The different analyses were used to study different crystalline structures of the deposited thin film, and estimate crystallite size. The crystallites size is reached by the Scherrer relation (Eq. 2.2), or the average crystallite size d can be estimated by considering the source wavelength (λ), FWHM of the peak (Δ), and the tilting angle (2θ); all adjusted by a correction factor (k), usually in the range of 0.9 [4].

$$d \simeq \frac{0.9\lambda}{\Delta(2\theta)\cos\theta} \qquad (2.2)$$

2.3.4 X-ray Reflectometry (XRR)

This technique has also been used to determine precisely the density and thickness of deposited thin films. In this case, the incident beam is launched with grazing incidence angle (θ) with the sample surface and a detector is used to measure the specular reflective intensity.

2.3.5 Atomic Force Microscopy (AFM)

The atomic force Microscopy (AFM), belongs to the family of near-field microscopes. This technique was developed in 1986 by Binning et al. The operation of the AFM is based on the interaction between a tip and the studied surface. The atomic force microscope (Fig. 2.4) operates at air, vacuum and also in liquid medium, which permits its use in various fields such as electrochemistry and biology. The sensitivity of its tip to different types of interaction, which can be controlled, allows in addition to surfaces imaging, measurements of several properties as: elasticity, viscosity, mechanical tribological (friction, adhesion), magnetic and electric, with a resolution of the order of nanometer and this for practically any type of sample.

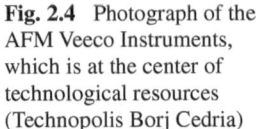

Fig. 2.4 Photograph of the AFM Veeco Instruments, which is at the center of technological resources (Technopolis Borj Cedria)

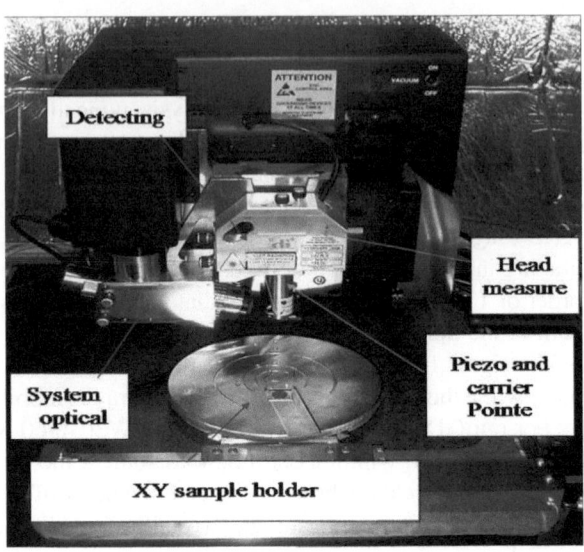

(a) **Equipment**

The AFM used (Fig. 2.4) to characterize our samples is an multimode Nanoscope III (Veeco Instruments) AFM. This device can be operated in 'contact' mode or "tapping" mode. The central element is the probe, it is a lever on which a nanometric tip is mounted. This tip is in mechanical contact with the surface to determine its morphology. Through a piezoelectric tube, the tip can be moved in X, Y and Z directions. The interaction between the tip and the surface is characterized by the measure of a physical data of the lever as: its deflection and amplitude of vibration. This measure is in most cases done by sending a laser beam at the back of the lever and detecting the reflected beam (Fig. 2.5). A control system of the tip maintains constant interaction strength, allowing to reconstruct point by point the topology of the surface. The regulation of the peak altitude is made by comparing the measured signal to the control signal. The error signal is then returned to the piezoelectric system and the peak altitude is then changed.

(b) **Different AFM operating modes**

«Contact» **mode**: when the tip and the sample is continuously in contact, the strength of interaction is responsible of the lever deflection on a quasi-static way. Most of the topographic images are obtained with this operating mode. The multimode AFM software contains algorithms that measure and present results as: cross-section analysis, roughness measurement, particle size analysis, depths analysis, the spectral density of the pattern, histogram analysis, forces measurement …

'Tapping' **mode**: tapping is the most used mode for the study of rough surfaces. In fact, it avoids degradation of the tip, friction and displacement of material from the surface of materials, since the interaction strength is

Fig. 2.5 Operating principle
of a atomic force microscope
(AFM)

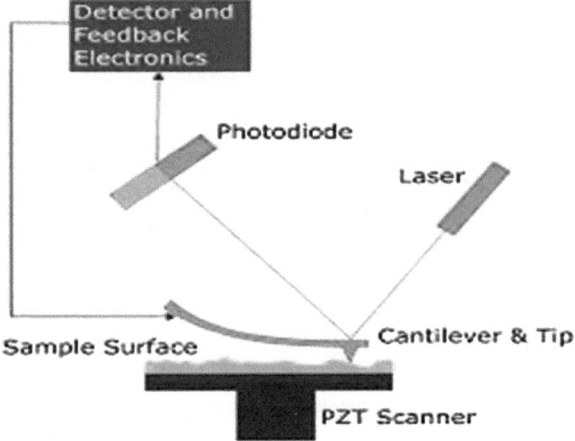

normal to the surface (no tangential component). Due to the high vibration frequencies of the tip that hits the surface of the sample intermittently, it also allows to minimize the effects of capillary action taking place in the presence of water film. For these reasons, we have used this mode for morphological studies of TiO_2 thin films.

2.3.6 Scanning Electron Microscope (SEM)

The scanning electron microscope (SEM) is used to get the microscopic images of a sample in vacuum with an electrons beam (typically between 5–25 keV energy). Under the effect of the electron, sample is the seat of a large number of physical phenomena (excitation of plasmons, emission of electrons …). In a conventional microscope, sensor and electron gun are in the same side of the sample. The most detected phenomena is the emission of secondary electrons. Those are the ejected electrons from the sample by the primary electrons beam having energy less than 50 eV. Backscattered electrons can also be detected. These are the primary beam electrons that have sufficiently deflected from their initial trajectory to sample out of elastic or quasi-elastic collisions, without energy transfer. Their energy is then so close to the primary beam energy. The achieved images represent essentially a topographic contrast. Indeed, the emission depth of secondary electrons is less than 10 nm. In addition, the number of emitted electrons depends on the 'shape' of the surface and the detected number depends on its orientation from the detector and the gun. Backscattered electron images are sensitive to the atomic number of the elements but the topographic contrast is less good. The scanning microscopes are often equipped with an energy dispersive spectrometer (EDX, "Energy Dispersive X-ray spectrometer"). In this work, the SEM analyses were performed at INRS-EMT by means of a JEOL JSM 7401F microscope.

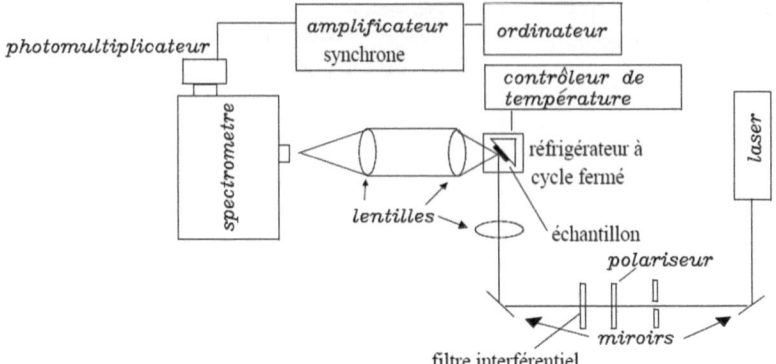

Fig. 2.6   Principle of operation of the assembly photoluminescence

2.3.7 Photoluminescence Setup

The experimental set-up is shown in Fig. 2.6. An ionized Argon laser ($\lambda = 488$ nm)
is used to excite the sample which is placed in vacuum. The light emitted from the
sample is focused by two lenses system at the entrance of a spectrometer equipped
with a GaAs photomultiplier cooled with water. The current given by the photomultiplier is converted to voltage (by a resistors box) which is measured using a
synchronized detection itself connected to a computer. The synchronized detection
used requires an excitation beam chopped at a fixed rate.

2.4 Electrical and Optoelectronic Characterizations

2.4.1 Test of Gas Sensors

The chosen sensor element is TiO$_2$ thin film of ~50 nm thick. In order to test
this sensor, gold electrodes closely spaced (3 mm) are deposited on this film by
evaporation. The TiO$_2$-based gas sensor is placed in a Chamber where variation of
conductivity is measured with different gas concentration, by controlling the temperature of the sample from ambient to 300 °C. The heating elements and the ther-
mocouple are connected to a temperature controller. Figure 2.7 shows the whole
experimental set-up used to test gas sensors. All measures devices are connected
to a computer via an 'Agilent Technologies' GPIB card that allows simultaneous
interfacing of up 14 devices. The measurements are controlled through programs
developed with the HP VEE programming language.

An HP 4140B pico-ammeter [5] is used for the conductivity measurement in static regime. HP4140B (Fig. 2.8) allows doing three types of measures
simultaneously:

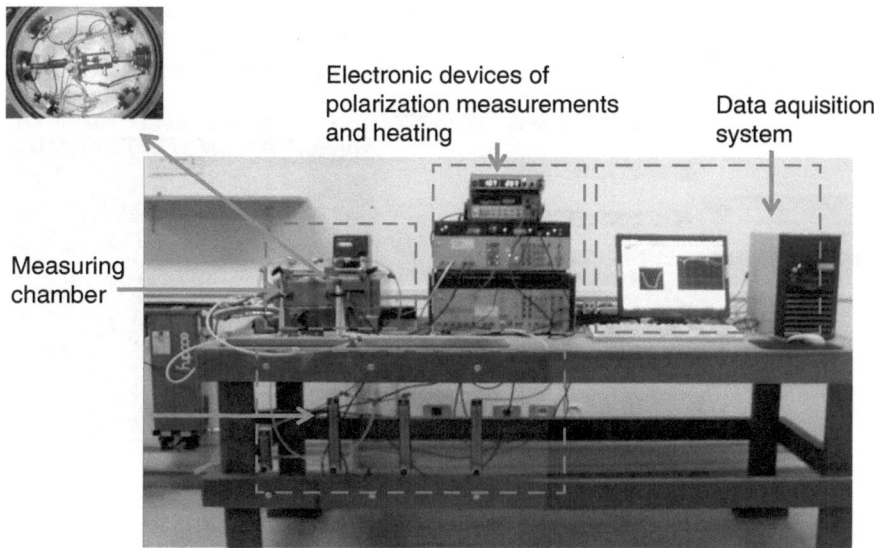

Fig. 2.7   Diagram of the tester to IPEST [8]

Fig. 2.8   Pico-ammeter
HP4140B

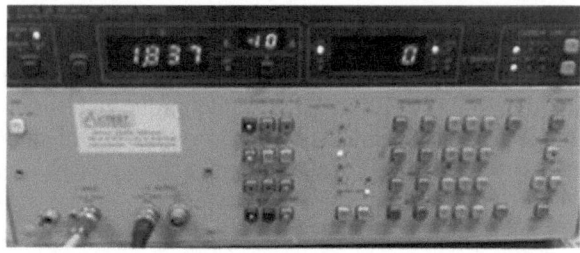

- conductance versus temperature: G (T).
- the conductance versus time at constant temperature: G (t),
- The current-voltage characteristics: I (V)

The sensors responses are visualized in real time. The conductance dependency on time and temperature changes was treated with the following procedures: isotherm and temperature cycles. An image of the program interface is shown in Fig. 2.9.

2.4.2   Light Beam Induced Current Technique (LBIC)

About 90 % of photovoltaic electricity is produced from mono and multicrystalline silicon cells. Knowledge of the diffusion length (or life time) of minority carriers in Silicon used in the development of solar cells is necessary in order to

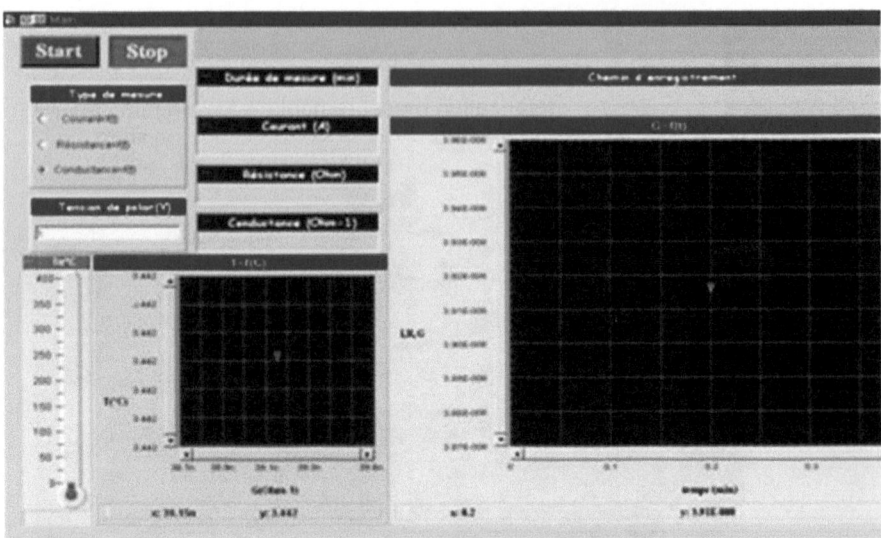

Fig. 2.9 Image of the interface of the program acquisition and temperature control based on time

adapt and optimize fabrication processes. The aim of the LBIC technique is the diffusion length measurement and the lifetime of minority carriers in mono and multicrystalline Silicon. LBIC technique measures the variation of the induced current at the terminals of a photovoltaic cell under monochromatic illumination (laser) focused at a point of a few µm in diameter. Minority carriers are photo-generated in a given point on the surface of the cell, producing the LBIC current. The LBIC current measurement enables to calculate the diffusion length of minority carriers. Figure 2.10 shows the LBIC set-up. The multicrystalline Silicon cell is mounted on motorized x y z stage. The x and y motors provide lateral movement of the sample compared to focused and fixed laser spot. The z motor is used to focus the laser on the diode surface. The sample is fixed on the x y z motorized stage with vacuum pump. The short-circuit current collected at the terminals of the cell is converted, amplified and filtered by a synchronized detection. At the output of the synchronized detection, the signal passes through a digital Voltmeter to a microcomputer using an IEEE cable. Once the value of the LBIC current registered, the microcomputer controls the motor in x (or in y) to change the position of the diode with the incident laser for a new LBIC measure. In this experiment, the red (He–Ne) laser of wavelength 628 nm is used.

The principle of measurement is illustrated in Fig. 2.11.

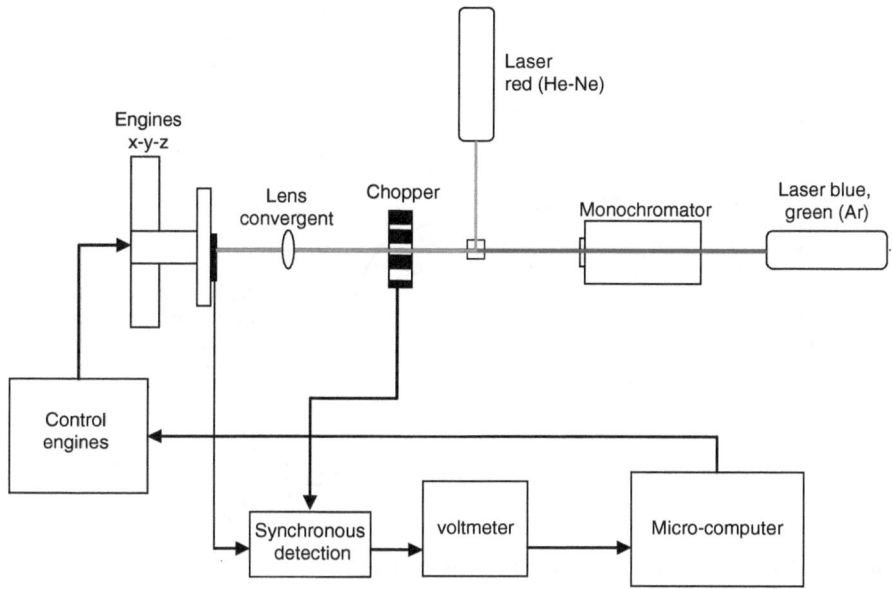

Fig. 2.10   Schematic assembly for measuring the current LBIC [9]

Fig. 2.11   A test structure
for measuring the current
in the case of LBIC cell
multicrystalline Si treated PS.
LBIC current measurement to
estimate the diffusion length
of minority carriers

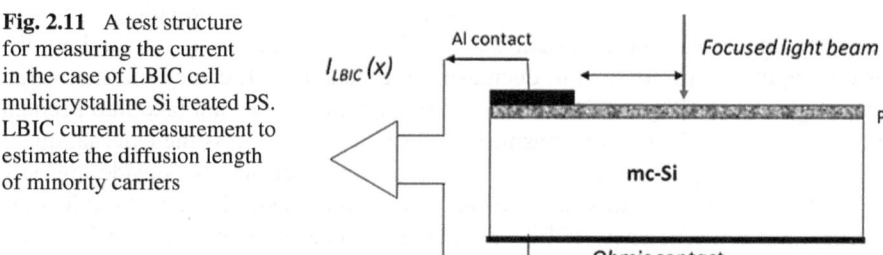

2.4.3 Quantum Efficiency

The quantum efficiency gives information on the overall performance of the
structure. The external quantum efficiency at a given energy is the ratio of the number of collected carriers on number of incident photons. If each photon of a specific
wavelength generates a collected charge carrier, the quantum efficiency is unity.
As it is noted earlier, the Silicon absorption coefficient varies with the wavelength
of the radiation. The photons with low wavelength are absorbed close to the surface (to area of the transmitter of the cell) while those with high wavelength are

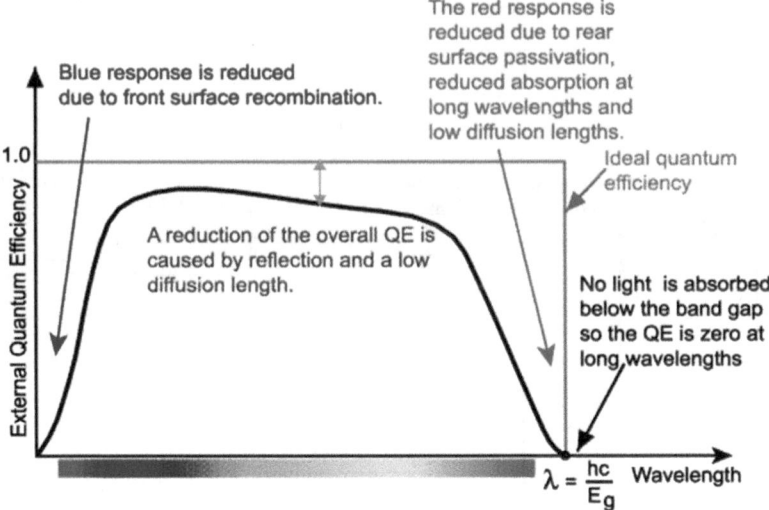

Fig. 2.12   External quantum efficiency of a solar cell [10]

absorbed more in depth into the device (at the level of the base). We can therefore access the spectral behaviour and the efficiency of each region constituting the cell. This parameter can be affected by several factors: the surface and bulk recombination phenomena play an important part. Reflection on the front and the low diffusion length of carriers lead to decreasing the quantum efficiency on the overall wavelengths range (Fig. 2.12). The presented parameter does not take into account optical losses as reflection or transmission through the cell: it is the external quantum efficiency (EQE). We can correct the quantum efficiency by considering only the absorbed photons, in this case we take into account losses by reflection: it is the internal quantum efficiency (IQE) which takes into account only the specific characteristics of the cell (diffusion length, surface and volume recombination).

2.5 Optical Characterisations

2.5.1 UV-Visible Spectrometer

UV-visible spectrophotometry is a characterization technique to reach the optical properties (transmission and reflectivity) of the materials. The measurements are performed on a wide wavelength range from 175 nm (UV) to 3,300 nm. The spectrometer is equipped with an integrating sphere which allows us to take into account the backscattered light (Fig. 2.13) [6]. The entire light incident in any direction to this highly reflective walls sphere is perpendicular reflected to detectors located on the wall of this sphere.

Fig. 2.13 Diagram of the integrating sphere used

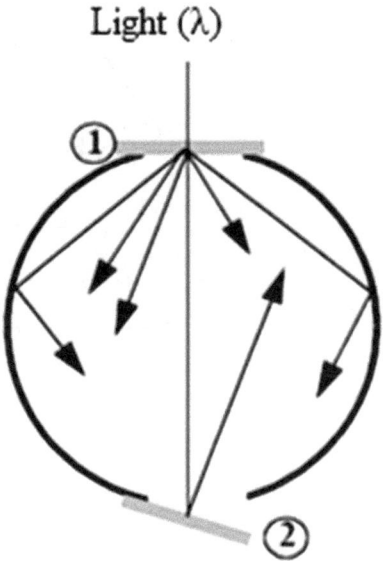

Light (λ)

The measurement of the total transmission (T) is performed by placing the sample at the entrance of the sphere (position (1), Fig. 2.13), the light that comes from a polychromatic source first penetrate the sample. The transmitted part is collected by a detector located on the surface of the sphere. The reflection is measured by placing the sample at position (2) (Fig. 2.13). The R(λ) and T(λ) measurement are performed using a lambda 950 spectrometer, equipped with an integrating sphere (150 mm or PELA 1021 infra gold) which allows to take into account the backscattered light.

2.5.2 Ellipsometry

Ellipsometry is an optical technique for surface analysis based on the measurement of the change in the state of polarization of the light after reflection on a flat surface. This technique, which the principle was discovered a century ago, knows a great success since 20 years thanks to the use of computing and electronic control of the motor, allowing optimization of measures.

(a) **Principle of measurement** [7]

Consider a plane wave arriving on a flat surface. Part of the wave is transmitted or absorbed through the surface, another part is reflected by the surface (Fig. 2.14). In this work, we consider only to the reflected part. The incident wave can be decomposed along two axes:

- one parallel to the incidence plane
- the other perpendicular to the incidence plane.

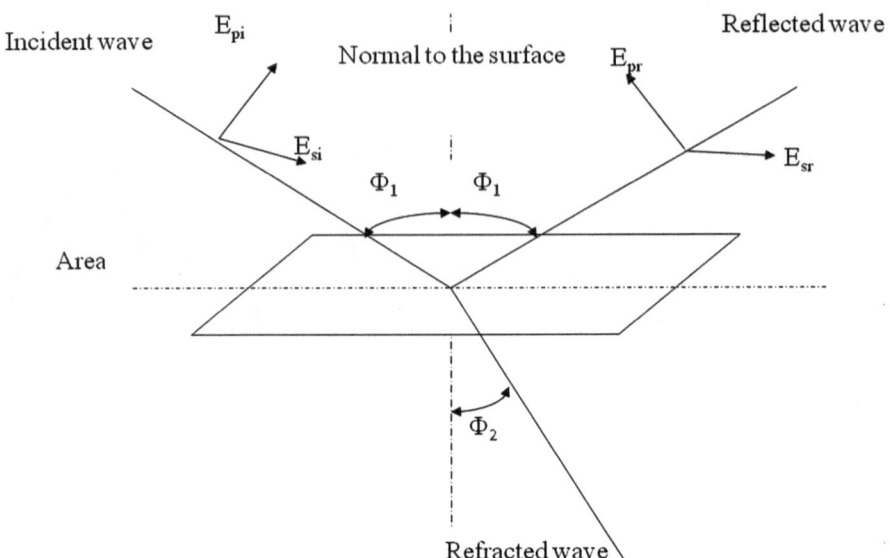

Fig. 2.14 Reflection of the polarization axes in the surface of the sample

The used index are: p for parallel, s for perpendicular, i for incident. The change of the electric field after reflection on the sample can be represented by two factors acting on each of the field components:

- the reflection coefficient of the sample for polarization parallel to the incidence plane is:

$$r_p = E_{pr}/E_{pi} = \left|r_p\right| \exp\left(j\delta_p\right) \qquad (2.3)$$

- the reflection coefficient of the sample for polarization perpendicular to the incidence plane:

$$r_s = E_{sr}/E_{si} = \left|r_s\right| \exp\left(j\delta_s\right) \qquad (2.4)$$

Both r_p and r_s coefficients are complex. Their modules $|r_p|$ and $|r_s|$ represent the variation of the field component amplitude, and their phases δp and δs, represent the delay introduced by the reflection. In practice, the measured is the ratio of these two factors, which is expressed in the following form:

$$r_p/r_s = \tan \psi \, \exp(j\Delta) = \rho \qquad (2.5)$$

ρ measurement leads to the identification of two quantities (ψ and Δ, or $\tan \psi$ and $\cos \Delta$). A measurement at a given wavelength and incidence angle will allow the calculation of the index n and k of a substrate knowing the thickness of the layer, or the index n and thickness e of a layer knowing extinction coefficient k. Finally,

Fig. 2.15 SOPRA GES5 spectroscopic ellipsometer

the previous relationships allow us to determine some optical characteristics such as refractive index, extinction coefficient and absorption coefficient.

(b) **Experimental setup**

The used device is a rotating polarizer ellipsometer (model GES5 de SOPRA, Fig. 2.15).

This Ellipsometer is composed of a source arm and detection arm that can be oriented by rotation on goniometer around an axis through the surface of the sample. The incidence angle is defined by the optical axis of the source arm and the normal to the surface of the sample.

Ellipsometry includes:

- **The source**: it is a short arc Xenon lamp at high pressure, of a very low residual polarization. It emits on the visible spectrum, from near ultraviolet to near infrared. The beam size (0.25 mm) from the arc, is collimated by a high focal length mirror to obtain a slightly divergent beam (0.3 mrad), which is necessary for a good definition of the incidence angle, and flatness of the incident wave.
- **The polarizer**: once collimated, the beam polarization is modulated by a rotating polarizer and low deviation with high extinction rates. It runs at a speed of 40 rounds per second.
- **The goniometer**: ellipsometry arms are mounted on a high-resolution double ring goniometer. The angular resolution is a key factor for the proper calibration of the instrument and its use in photometry.
- **Sample holder**: the sample is maintained by suction, (allows to limit all constraints on the surface of the holder), and adjustable following two directions. The orientation of the optical surface can be spotted using an autocollimator. The surface of the sample must be free of dust and oily traces.
- **Analyzer**: positioned after the sample, it is fixed during the measurement. After passing through the Analyzer, the beam is focused at the input of an optical fiber for the optical connection with the spectrometer.

- **The spectrometer**: it is mechanically independent from rotation block, for better handling. The beam at the output of fibre is focused on the slot of a Czerny-Turner spectrometer mirror of 500 mm focal length. The spectrometer consists of two additives dispersive elements separated by a fixed intermediate slit. The first part dispersive element is a prism dispersing spectrum on the intermediate slot. It acts as a filter of linewidth defined by the slot width and a center wavelength variable depending on the position of the Prism. The isolated spectral bandwidth is then dispersed in the second part through a higher order grating, allowing a high spectral resolution. The Prism ensures the selection of the grating order.
- **The photomultiplier**: the detector is a photomultiplier working from the ultraviolet to the near infrared.

Finally the spectroscopic ellipsometry found its applications in areas involving thin films or surface analysis such as:

- Optics: characterization of thickness and refractive index of dielectrics and metals used in the fabrication of mirrors, anti-reflective layers and polarizing surfaces;
- Solid-state physics: measurement of permittivities which characterize crystallinity or the amorphization state of a material, semiconductor element band structure, or the coating of a deposited layer (polymer), porosity (by determining the concentration of the vacuum in the material).

2.6 Conclusion

In the first part of this chapter we have presented the synthesis method of titanium dioxide layers. In addition and in order to facilitate the identification of TiO$_2$ thin films, we presented the main techniques to perform electrical, structural, morphological and optical characterizations of these films. The second part was devoted to the description of a gas sensor set-up and to the principles of optical and optoelectronic measurements.

References

1. Saminadayar K (1997) Cours de Physique des couches minces, Formation doctorale Microélectronique
2. Chatterjee S (2008) Titania-germanium nanocomposite as a photovoltaic material. Sol Energy 82:95
3. Brundle CR, Evans CA, Wilson S (1992) Encyclopedia of materials characterization—surfaces, interfaces, thin films. Elsevier, Amsterdam, p 800
4. Patterson AL (1939) The scherrer formula for x-ray particle size determination. Phys Rev 56:978
5. Bejoui A (2009) Etudes électriques, structurales et morphologiques des oxydes de cuivre (Cu$_2$O, CuO) pour l'application capteur. Mastère, Institut National des Sciences appliquées et de Technologie

6. Selvan JAA (1998) Thèse docorat, Université de Neuchâtel. ISBN 3-930803-60-7
7. Adachi A (1982) J Appl Phys 53:8775
8. Labidi A, Bejaoui A, Ouali H, Akkari FC, Hajjaji A, Gaidi M, Kanzari M, Bessais B, Maaref M (2011) Appl Surf Sci 257:9941
9. Dimassi W (2007) Etudes spectroscopiques et optoélectroniques de la passivation des cellules solaires au silicium multicristallin basée sur l'élaboration du silicium poreux. Thèse doctorat, Faculté de Sciences de Tunis
10. Honsberg C, Bowden S (1998) Photovoltaics: devices, systems and applications (CDROM). University of New South Wales, Sydney

Chapter 3
Microstructure and Optical Properties of Pure and Cr-Doped TiO_2 Thin Films

Abstract This chapter describes the structural, optical and electrical properties of TiO_2 layers grown on various substrates such as quartz, intrinsic silicon and both P and N types silicon. This chapter starts with a brief description of the deposition parameters. In particular, the influence of chromium content as well as the growth and mechanism conditions on the structure, morphology and optical proprieties (size of crystallites, mesh parameter, morphology, chemical composition, optical index, bandgap energy, reflectivity, etc.) of the final prepared films are described.

Keywords X-ray diffraction (XRD) · X-ray photospectroscopy (XPS) · X-ray reflectometry · Fourier transform infra-red spectroscopy (FTIR) · Raman spectroscopy · Ellipsometry · UV-visible spectroscopy · Band gap energy

3.1 Introduction

This chapter is dedicated to the study of the structural, optical and electrical properties of TiO_2 layers on a substrate (quartz, intrinsic silicon, P type silicon and N type silicon). A parametric study was undertaken in order to get a fine control of different films properties. Thus, the influence of deposition parameters (Cr doping, effect of deposition temperature, development time, power) was closely correlated with the microstructural and optical characteristics (size of crystallites, mesh parameter, morphology, chemical composition, optical index, bandgap energy, reflectivity, etc.). TiO_2 thin films doped with Cr were developed by co-deposition using magnetron sputtering process with two targets, the TiO_2 and the chromium ones. Index of purity are 99.995 and 99.99 % respectively for Cr and Titanium oxide. The applied power to the target, the pressure, flow and volumetric ratio of Ar and O_2 gases, as well as the possible polarization of the substrate were chosen based on previous work (see Table 3.1) [1].

© The Author(s) 2015

A. Hajjaji et al., *Chromium Doped TiO2 Sputtered Thin Films*,
SpringerBriefs in Manufacturing and Surface Engineering,
DOI 10.1007/978-3-319-13353-9_3

Table 3.1 Deposition conditions used to prepare thin films of TiO_2

Puissance appliquée à la cible de TiO_2 (W)	Puissance appliquée à la cible de Cr (W)	Débit Ar (sccm)	Débit O_2 (sccm)	Pession (mTorr)
360	8	80 %	20 %	1.2
//	15	//	//	//
//	25	//	//	//
//	50	//	//	//
//	100	//	//	//
//	150	//	//	//

3.2 Effect of Cr Concentration

3.2.1 Structural Properties

3.2.1.1 XPS Analysis of TiO_2 Films

XPS spectra of the elements Cr, Ti, O of Cr doped thin films are obtained from high resolution windows of 20 eV width after etching for 600 s (approximately 15 nm in depth from the surface). Cr atomic percentage content according to the applied power to the Cr target is shown in Table 3.2.

Total Cr concentration increases as with the applied power to the Chrome electrode. It varies from 7 to 17 % for power values from 25 to 150 W. In XPS spectra, there are clearly two components at 577.05 and 585.8 eV attributed respectively to Cr 2p3/2 and Cr 2p1/2. Low energy peak suggests the presence of the metallic phase of chromium so that high-energy peaks suggest an oxidized State of CrO_2 and Cr_2O_3 of chrome oxides as well as Cr–Ti binding types. These chromium and titanium states vary according to the applied power.

Figure 3.1 shows the experimental measurements and theoretical adjustments obtained for a TiO_2–Cr sample (P_{Cr} = 25 W). The peaks of the oxidized Cr (CrO_2) and (Cr_2O_3) appear at energies 581.03 and 584.1 eV for the 2p3/2 state and 590.83 and 593.9 eV for the 2p1/2 state (ΔE = 9.8 eV) with respective concentrations of 58 and 23 % for the Cr_2O_3 and CrO_2 states (Fig. 3.1a). Peaks of metallic Cr appear at energy 578.06 and 587.2 eV (ΔE = 9.2 eV) and its concentration is estimated at 16 %. For the Cr–Ti bindings, 2p1/2 and 2p3/2 degeneracy appear around 583.77 and 592.97 eV (ΔE = 9.2 eV). The concentration is of the order of 7 %. The Ti2p

Table 3.2 Atomic concentrations depending on the applied power

Puissance (W)	Cr-metalic (at.%)	CrO_2 (at.%)	Cr_2O_3 (at.%)	Cr–Ti (at.%)	Cr total absolute (at.%)
25	1.12	4.06	1.61	0.21	7
50	0.72	4	2.24	1.04	8
100	3.12	5.2	2.6	2.08	13
150	1.87	11.73	3.06	0.34	17

Fig. 3.1 High resolution
XPS spectra measured
for a sample of TiO_2–Cr
(P_{Cr} = 25 W) **a** Cr2p; **b** Ti2p;
and **c** O1s

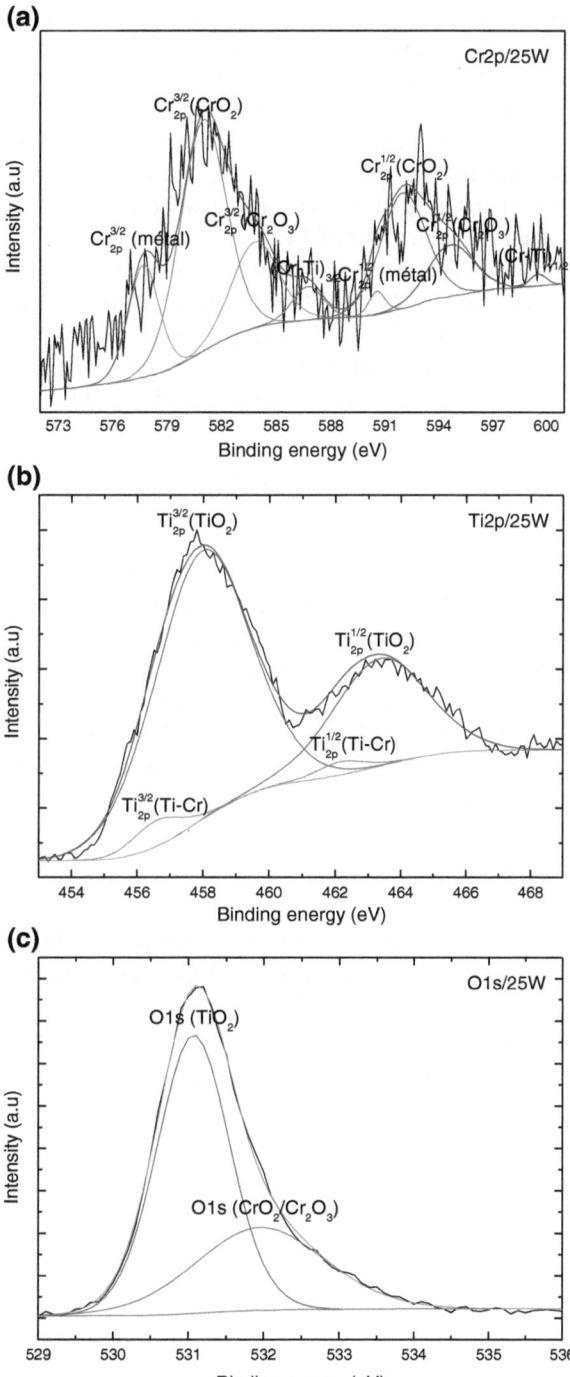

lines (Fig. 3.1b) indicate the presence of an oxidized phase (TiO_2) to 457.93 eV (p3/2) and 463.89 eV (p1/2) ($\Delta E = 4.54$ eV) with a concentration of approximately 94 %. A low concentration of Ti–Cr binding appear at the energies 454.25 eV (p3/2) and 466.99 eV (p1/2) ($\Delta E = 9.2$ eV), its concentration is about 7 %.

Bonds that involve oxygen are of two types (Fig. 3.2c). Those involved in the TiO_2 are located at 531.1 eV, and those involved in the Cr oxides (CrO_2 and Cr_2O_3) are located at 531.9 eV. The concentration of the first is 65.25 % and the second is in the order of 34.75 %.

These XPS results shows indeed that the chrome is present in two forms, namely the Cr–O and Ti–Cr. Chromic oxide appears in two States: the CrO_2 (majority) and the Cr_2O_3. High resolution XPS analysis gives information on the nature of the different phases formed in the TiO_2 layers which are mainly the Cr_2O_3, CrO_2, TiO_2, the metallic Cr and Cr–Ti bindings.

3.2.1.2 Analysis by X-ray Diffraction (XRD)

Figure 3.2 shows the X-ray diffraction diffractograms in grazing incidence of Cr-doped TiO_2 thin films deposited on intrinsic Si substrate at room temperature. We note that these layers are semi-amorphes and there is a smoothly improvement in the microstructure depending on the doping concentration.

The two peaks observed at $54° < 2\theta < 56°$ may be corresponding to a mixture of anatase and rutile phases. The appearance of these two peaks indicates a beginning of growth despite the low deposition temperature ($T_d = 25$ °C). The interaction of the substrate with the plasma probably leads to a local heating with temperature that could reach 100 °C [2].

3.2.1.3 Analysis of the Film Density by X-ray Reflectometry

Figure 3.3 illustrates changes in semi-logarithmique scale of the X-ray intensity reflected by the intrinsic TiO_2–Cr/Si structure depending on the doping

Fig. 3.2 XRD diffraction (grazing incidence) TiO_2 layers function of the atomic percentage of Cr

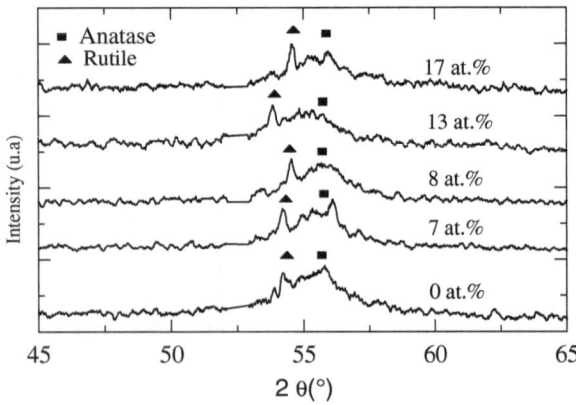

concentration with Cr. Note that the thickness of the deposited layer can be determined from the angular positions of the Kiessing fringes (Fig. 3.3). An estimate of the thickness d was determined using the relation:

$$d = \frac{\lambda}{2\Delta\theta}[2, 3] \qquad (3.1)$$

with 2θ is the angular interfringe and λ the radiation wavelength.

Figure 3.4 shows the X-reflectivity diagrams for different TiO$_2$ films doped with Cr. Reflectance curves reveal a single level of total reflection, confirming that the density of the upper layer (TiO$_2$) is higher than the substrate (silicon) and the homogeneous character of these films. The measurement of the total reflection critical angle is used to determine the electron density of the layers and their mass density [3].

The critical angle θc is defined from the moment where the reflected intensity decreases brutally. Using this value, the thin film density can then be calculated. It is possible to determine the critical angle using software (Parratt) to adjust the reflectivity

Fig. 3.3 Typical diagram X reflectivity of a film of TiO$_2$ doped Cr (2 at.%)

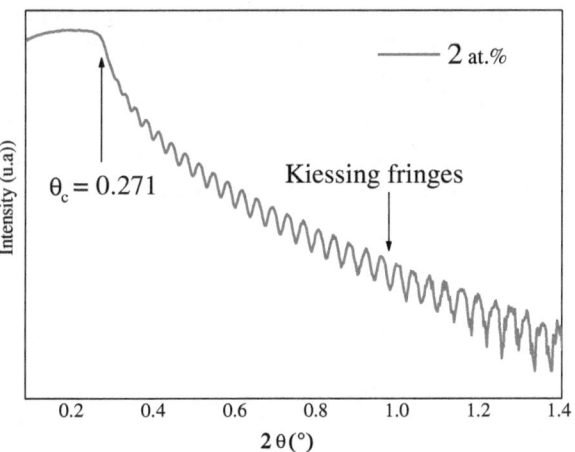

Fig. 3.4 Evolution diagrams reflectivity X TiO$_2$ layers according to Cr content

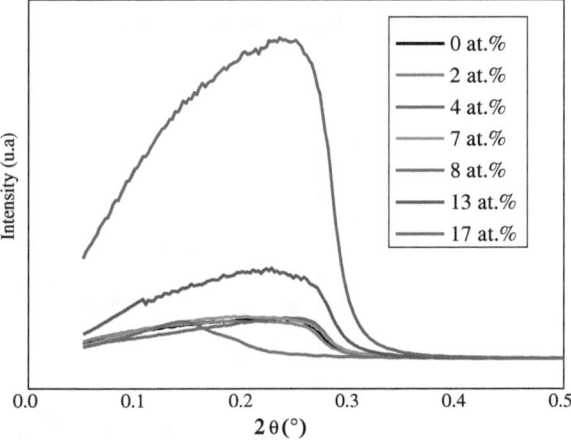

XRD diagram. However, it is not necessary to use this software to find the critical angle. Using a calculation software computes the derivative of the curve to obtain a minimum at the inflection point of the intensity. This minimum will define the critical angle θc. By assuming a few approximations we can use the following formula:

$$\vartheta_c = \sqrt{2\delta} = \sqrt{\frac{r_0 \lambda N_A \sum (f1)}{\pi A} \rho} \tag{3.2}$$

where:

r_0 $2{,}813 \times 10^{-6}$ nm is the Bohr radius
(f1) atomic form factor (e-/at)
M   molar mass (g/mol) of the thin-layer element
N_A 6.02×10^{23} (at/mol) = number of Avogadro
λ 1.541 Å Cukα copper radiation and θc = critical angle (rad) measured by
      dI/dθ or determined from δ during the fitting. The density value can be
      directly obtained from the critical angle θc, as well as the molar mass M and
      the atomic form factor ({f1}). Typical parameters used for density calculation
      are shown in Table 3.3.

Thus, the calculated density varies between 3.75 g/cm^3 for the pure TiO$_2$, and 4.03 g/cm^3 for TiO$_2$–Cr (at 17 %). Figure 3.5 presents the evolution of the density with the incorporation of the Cr element.

Table 3.3 Typical film density calculation parameters

Sample	θ_c (deg)	θ_c (rad)	δ	ρ_e (e-/m^3)	masse molaire (g/mol)	Σ (f1) (e-/at)	ρ (g/cm^3)
Si	0.223	0.0039	7.57×10^{-6}	$7.12 \times 10^{+29}$	28.06	14.00	2.37
TiO$_2$	0.274	0.0047	1.12×10^{-5}	$1.05 \times 10^{+30}$	80	$3.8 \times 10^{+1}$	3.68

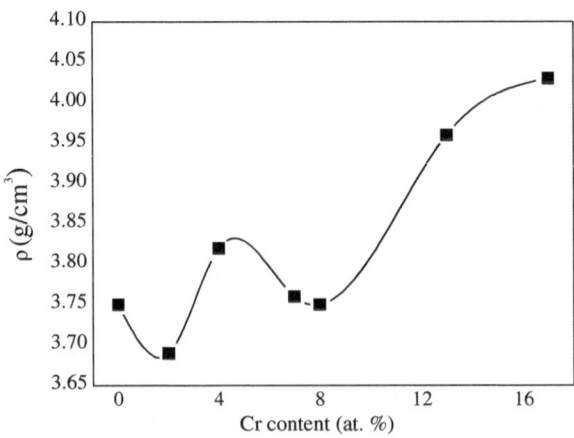

Fig. 3.5 Film density variation in terms of Cr content

3.2.1.4 Analysis by Fourier Transform Infra-Red Spectroscopy (FTIR)

FTIR method aim to analyze surface layers binding and study the influence of the doping elements presence on the nature of the adsorbed molecules. These measurements were made under vacuum at ambient temperature. Figure 3.6 shows the IR spectra of TiO_2 layers doped with Cr. Spectra exhibit mainly large and intense absorption band between 200 and 600 cm^{-1}. These bands are assigned to Ti–O–Ti vibrations type [4].

3.2.1.5 Raman Spectroscopy Analysis

Cr-doped TiO_2 thin films deposited on a Si substrate at room ambient temperature are analyzed by Raman spectroscopy technique in a frequency range between 100 and 850 cm^{-1} with a time of 15 s. The illustrated Raman spectra in Fig. 3.7 are consistent with XRD spectra. They show that TiO_2 layers performed at room temperature are in a semi-amorphe phase.

Fig. 3.6 FTIR Spectra of TiO_2 thin films as a function of Cr doping

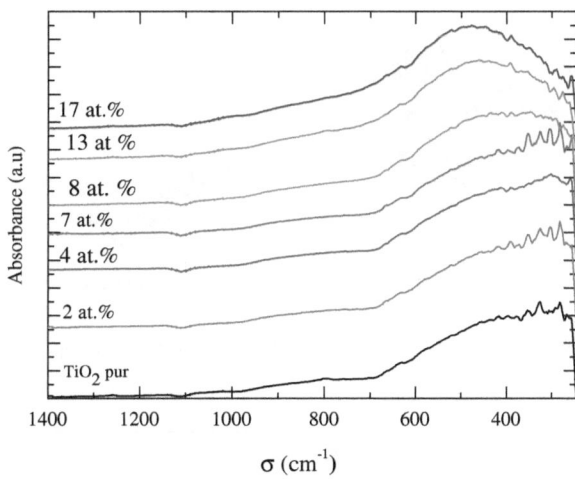

Fig. 3.7 Raman spectra of TiO_2 thin films with Cr doping

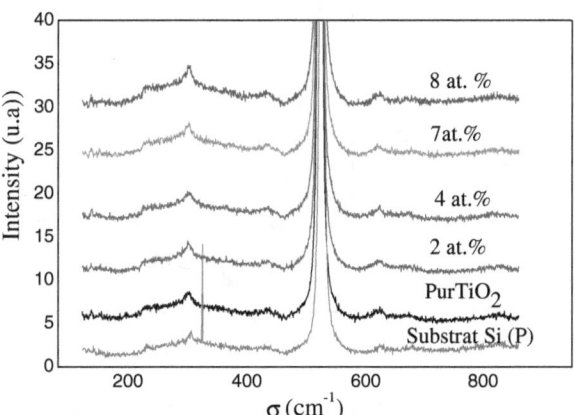

3.2.1.6 Films Morphology: Analysis by Atomic Force Microscopy

The AFM mirographs are carried out using the "tapping" mode. Figure 3.8 shows that the films exhibit a Nanoporous structure. Surface roughness (RMS) of Cr–TiO$_2$ films decreases with Cr concentration (Fig. 3.9) which could be explained by the densification of the film.

3.2.1.7 Observation by Scanning Electron Microscopy (SEM)

Cross-sectional SEM observations show that the TiO$_2$ film doped with Cr (17 %) has a very dense appearance compared to that obtained without doping (Fig. 3.10) and no cracks and voids have been observed. You can also notice the beginning of a granular appearance in the surface of the doped layer. The film thickness is between 120 nm (for pur TiO$_2$) and 200 nm (Cr–TiO$_2$ at 17 %).

Pure TiO$_2$ 4 at.% Cr   8 at.% Cr

Fig. 3.8 3D Migrographs of TiO$_2$ thin films with Cr doping

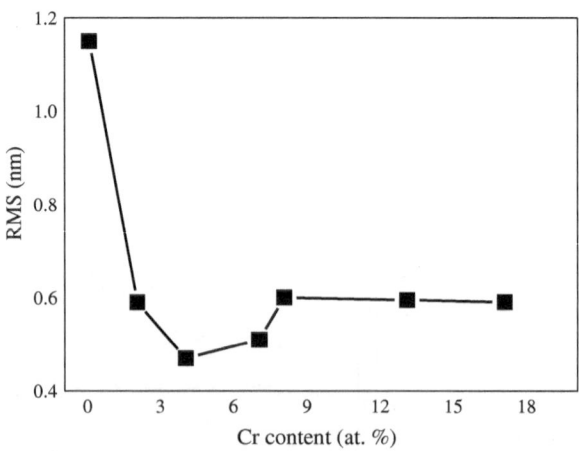

Fig. 3.9 RMS of TiO$_2$ films deposited at 25 °C as a function of Cr concentration

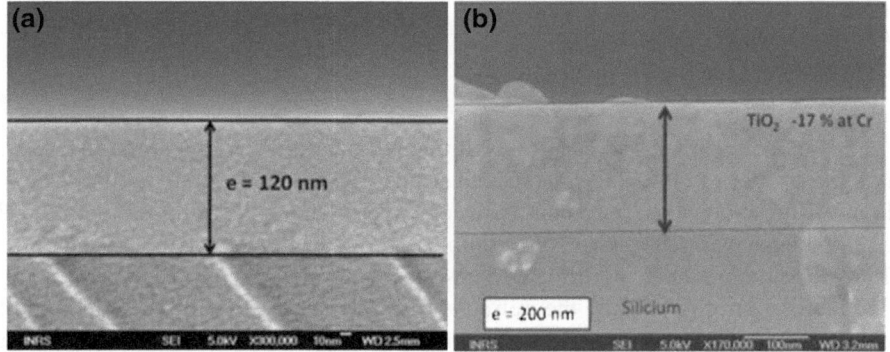

Fig. 3.10 Typical cross-section SEM micrographs of TiO$_2$ films (**a**, **b**) as deposited at 25 °C. **a** Pur TiO$_2$, **b** TiO$_2$–Cr (17 at.%)

3.2.2 Optical Properties

3.2.2.1 Optical Analysis by Ellipsometry

The interpretation of optical measurements requires a parametrization of the optical properties depending on the energy of the incident photon. Jellison and Modine [5] have developed a model of dielectric function for amorphous materials. In the near-infrared to near-UV domain, the dielectric function of an amorphous semiconductor can be described by a dispersion law based on the classical Lorentz oscillator and the density of states of Tauc [6]. In this model, the dependency of the dielectric constant on energy (E) in the dielectric function ε is given by:

$$\varepsilon_{2\pi}(E) = \begin{cases} \dfrac{AE_0C(E-E_g)^2}{\left(E^2 - E_0^2\right)^2 + C^2E^2} \cdot \dfrac{1}{E}; & E > E_g \\ 0; & E < E_g \end{cases} \tag{3.3}$$

$$\varepsilon_{1\pi}(E) = \varepsilon_\infty + \frac{2}{\pi} \cdot P \cdot \int_{E_g}^{\infty} \frac{\xi.\epsilon_2(\xi)}{\xi^2 - E^2} d\xi \tag{3.4}$$

where P is the principal part of the integral. Four of this law dispersion parameters are used to calculate the optical constants of the studied material [5]: the optical gap Eg, the pre-factor A containing the elements of the optical transitions matrix related to the material density, the energy gap between the middle of the valence band and the conduction band E$_0$ and the broadening parameter C. For example, Table 3.3 summarizes the model parameters of Tauc–Lorentz [7] used to analyze ellipsometry measures obtained for pure TiO$_2$ layers and Cr doped film with 7 % (see Table 3.4).

Table 3.4 Parameters of Tauc–Lorentz model used to fill experimental data

	A (eV)	B (eV)	C (eV)	Eg (eV)
TiO$_2$ pure	329	4.26	2.2	3.31
7 at.% of Cr	207	4.75	2.47	3.07

The model adapted for TiO$_2$ films structure consists of 3 layers: the first layer is the intrinsic Si substrate, the second is the TiO$_2$, and the third is the vacum. Figure 3.11 shows the agreement between experimental spectrum and theoretical one obtained with TiO$_2$–Cr (7 %) sample. The good quality of fit indicates that Tauc–Lorentz model is appropriate to describe the optical properties of such films.

Figure 3.12 shows that the refractive index increases with the amount of embedded Cr, the value of the gap energy decreases as the concentration of Cr increases; it goes from 3.31 eV the pure TiO$_2$ to 1.5 eV for the TiO$_2$–Cr (17 %) [8]. This result is so required for photovoltaic applications.

Indeed, the control of the bandgap value enable the covering of the useful solar spectrum and therefore improvement of the energy efficiency. The film thickness is between 110 nm (for pure TiO$_2$) and 190 nm (for film doped at 17 at.% of Cr).

Fig. 3.11 Typical ellipsometry spectra and best fit of TiO$_2$ doped Cr (17 at.%). **a** Psi, **b** Delta

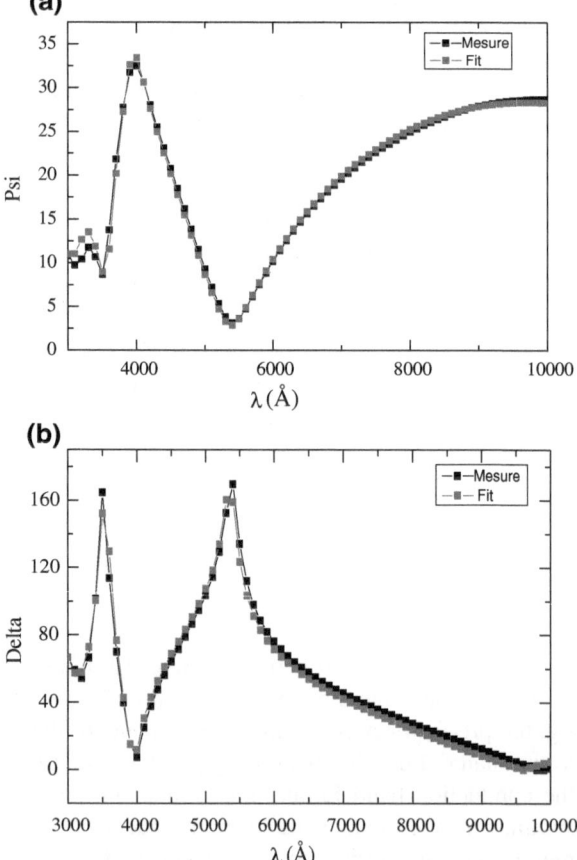

Fig. 3.12 Optical
properties and thickness
of TiO$_2$ film as a function
of Cr concentration. **a** The
refractive index, **b** band gap
and **c** thickness

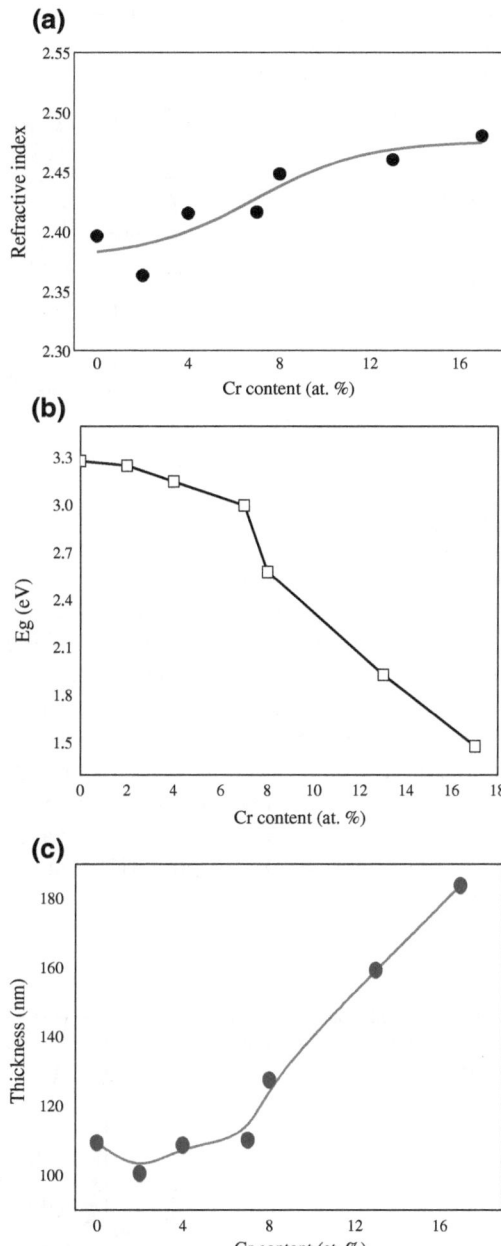

3.2.2.2 Analysis by UV-Visible Spectroscopy

Figure 3.13 shows the transmission spectrum of TiO$_2$ thin films, deposited on
quartz substrates at room temperature. This spectrum indicates that TiO$_2$ thin films
are transparent in the visible, and shows a red-shift with Cr incorporation. This

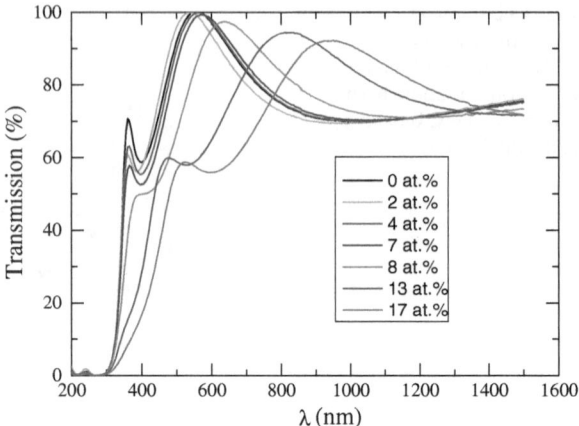

Fig. 3.13 Transmission spectra of TiO$_2$ film as a function of Cr concentration

shift is consistent with ellipsometry measurement of the optical band gap. The oscillations observed in the transmission spectrum T(λ) are due to interference phenomena and depend on the refractive index of the studied layer. The analysis of this kind of spectra leads to the thickness (e), index (n), the extinction coefficient (k) and the absorption coefficient (α) of the layer. The absence of interference fringes is consequence of the relatively perturbed surface of Cr doped TiO$_2$ films.

On the other hand, Fig. 3.14 shows the reflectivity spectra of pure and Cr-doped TiO$_2$ layers deposited on a Si P type substrate. There is a decrease of the reflectivity compared to the substrate, which shows that pure and Cr-doped TiO$_2$ films can play the role of antireflective layers in photovoltaic cells. The reflectivity of TiO$_2$–Cr films (17 %) shows a minimum at wavelength $\lambda = 600$ nm. It decreased by 44 % (substrate) 40 % (for pure TiO$_2$) and 9 % for the highly doped TiO$_2$ layers (17 at.%).

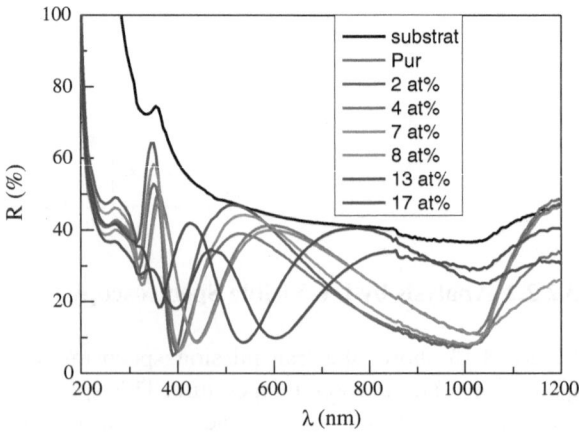

Fig. 3.14 Reflectance spectra of Cr doped TiO$_2$ thin films

3.3   Annealing Effect

In this section, the electrical, optical and microstructural characterization results performed on pure TiO$_2$ thin films and Cr-doped ones have been presented. These films have undergone an annealing at 550 °C for 60 min in a 3 areas oven under an oxygen flux. Analyses by X-ray diffraction (XRD) and Atomic Force Microscopy (AFM), are designed to analyze the effect of annealing on the morphology of the layers (size, mesh parameter, roughness).

3.3.1   Structural Properties

3.3.1.1   X-ray Diffraction

The effect of annealing under oxygen on the crystallinity of the TiO$_2$–Cr layers is shown in Fig. 3.15. The XRD patterns show the layers crystallization. The anatase phase predominates for Cr concentration ≤4 %. Beyond this value, a rutile transition phase is then observed.

The grain size was evaluated from full width at half maximum (FWHM) Δ of the X-ray diffraction peak using the Debye Scherrer formula [9]:

$$L = 0.9\lambda/\Delta \cos\theta \qquad (3.5)$$

where λ is the X-ray beam wavelength and θ the incidence angle.

Broadening due to the equipment can be corrected by taking as FWHM value: $\Delta' = \Delta - \Delta_0$ where Δ_0 is the measured value on a sample for which the theoretical linewidth is zero. It is therefore advantageous for the measurement of the grains size to consider the diffracted peaks at low angles ($\theta \leq 20°$), which would give for the standard linewidth $\Delta\theta \approx 0$ [10].

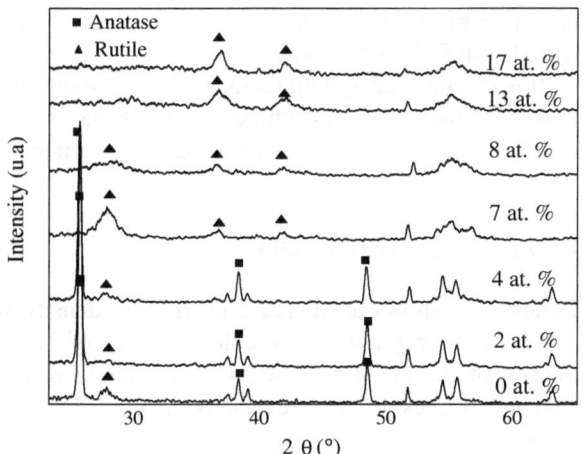

Fig. 3.15   XRD patterns of TiO$_2$ films as a function of Cr concentration (at.%)

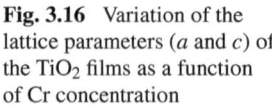

Fig. 3.16 Variation of the lattice parameters (*a* and *c*) of the TiO$_2$ films as a function of Cr concentration

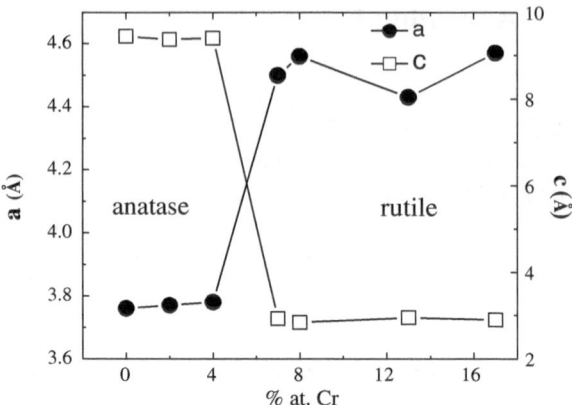

The size was estimated using (101) orientation for the Cr-doped TiO$_2$ films which crystallize in the anatase phase and (110) orientation for the rutile phase. When the doping concentration increases from 0 to 17 at.%, crystallite size varies between 35 and 6 nm. This suggests that the growth of TiO$_2$ grains depend on the presence of Cr metal. The lattice parameters (a, c) were calculated using the Bragg law from the (101) peaks for the anatase phase and (110) peaks for the rutile phase. Figure 3.16 shows the evolution of the a and c lattice parameters with the Cr concentration. The evolution of these two parameters shows a transition from anatase phase to rutile phase for Cr concentration of 7 %. This anatase-rutile phase transition is probably related to an increase in the concentration of metallic dopants (donor or acceptor). Indeed, acceptors of type Cr^{3+}, for example, introduce oxygen vacancies and thus accelerates the anatase-rutile transition [11, 12]. In such films, chromium oxide appears to be in two States, namely CrO$_2$ (the main component, which is also the most stable) and Cr$_2$O$_3$.

Cr^{3+} radius (0.62 Å) is similar to that of Ti^{4+} (0.68 Å), suggesting that it is easy to replace Ti^{4+} by Cr^{3+}. Similar behavior has been reported for TiO$_2$ films doped with Fe^{3+} ions [13].

It is synthesized and marketed in 1941 [14] at high temperatures, about 800 °C, thus our study aims to achieve rutile phase at low temperature as in our case (TA − R = 550 °C) and this due to doping elements as Cr, Mn, Fe.

3.3.1.2 X-ray Reflectometry

Figure 3.17 shows an increase in the film density with the atomic concentration that could be explained by the densification of TiO$_2$ films with the increase in the density of the metal particles. Cr particles tend to fill the pores within the TiO$_2$ film and as a result increase density.

Fig. 3.17 Effect of Cr
concentration on the film
density of the TiO₂ films as a
function of Cr concentration:
annealed at 550 °C

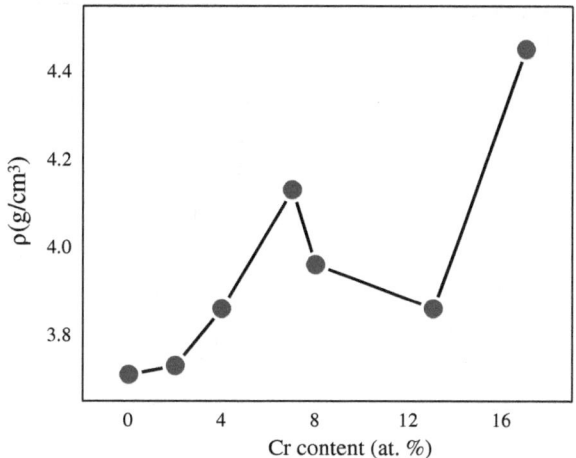

The calculated density ranges indeed from 3.71 g/cm³ for the pure TiO₂ and
4.45 g/cm³ for TiO₂–Cr (17 %).

3.3.1.3 Analysis by Fourier Transform Infrared Spectroscopy (FTIR)

Figure 3.18 shows the FTIR spectra of TiO₂ layers annealed at 550 °C,
depending on the Cr concentration. Below a 7 % of doping concentration, FTIR
spectra show the presence of two peaks located at 260 and 433 cm⁻¹, due to
the vibrations of Ti–O–Ti bonds corresponding to the anatase phase [4]. These
peaks disappear completely when the Cr concentration exceeds 7 % Cr and
are replaced by absorption peaks assigned to the rutile phase, located around
502 cm⁻¹ [4].

Fig. 3.18 FTIR spectra
of TiO₂ films annealed at
550 °C as a function of Cr
content

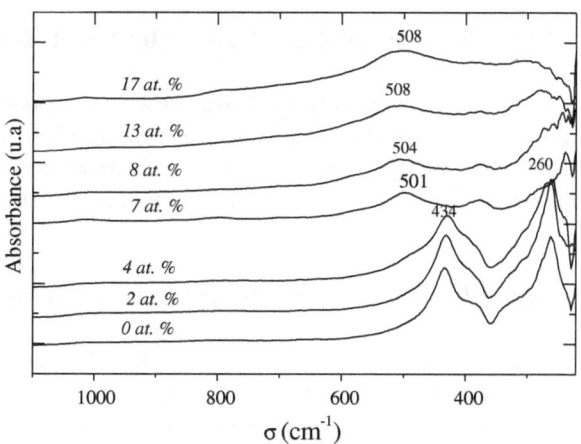

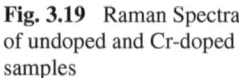

Fig. 3.19 Raman Spectra of undoped and Cr-doped samples

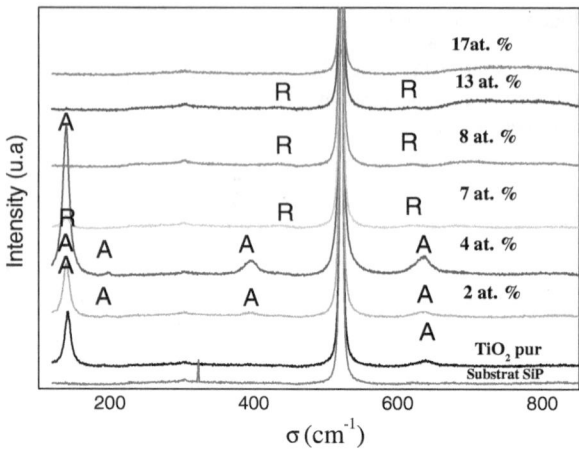

3.3.1.4 Analysis by Raman Spectroscopy

The analysis of Raman Spectra confirms the results obtained from the XRD and FTIR analysis (Fig. 3.19). The TiO_2 layers annealed at a temperature 550 °C show a good crystallinity, anatase-rutile phase change, was also confirmed for the layers having a Cr concentration exceeding 4 %.

In literature, the band position corresponding to the anatase phase is $Eg_A = 144$ cm^{-1} while that corresponding to the rutile phase is $Eg_R = 143$ cm^{-1}. We were interested to the EgA band because it is more intense and sharper, which facilitates observation and comparison with other Cr doped films and the phase shift control. Figure 3.19 shows the approximate presence of band located at 137 cm^{-1} for pure and Cr doped TiO_2 films. This band shift to lower frequencies and explained by several authors [15] as consequence of a small crystallite size (less than 15 nm).

3.3.1.5 Films Morphology: Analysis by Atomic Force Microscopy

2D AFM micrographs (Fig. 3.20), show also a phase change from the anatase to rutile from as we exceed a Cr concentration of 7 %. In fact, the surface roughness (RMS) of Cr–TiO_2 films increases by increasing Cr concentration and reaches a maximum for a concentration of 7 % (Fig. 3.21) [16].

3.3.1.6 Observation by Scanning Electron Microscopy (SEM)

The SEM images (Fig. 3.22) show a layered and dense structure of films with anatase phase. In contrast to the high Cr concentrations (rutile phase), films are then formed by nano randomly oriented grains.

Fig. 3.20 AFM (2D) scans of TiO$_2$ films annealed at 550 °C as a function of Cr concentration

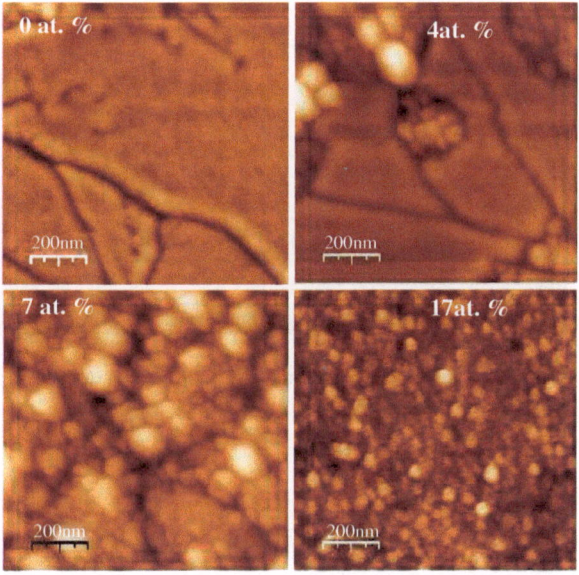

Fig. 3.21 Rms of TiO$_2$ films annealed at 550 °C as a function of Cr concentration

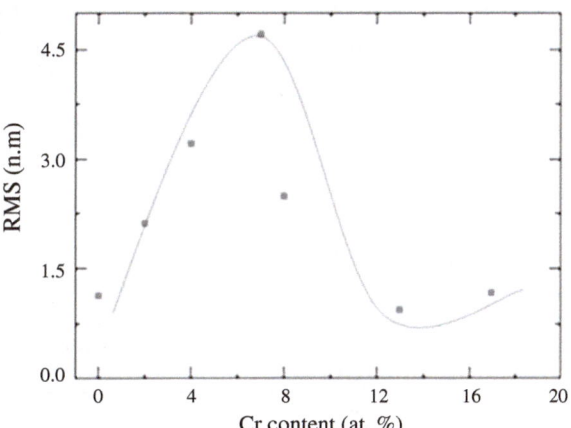

3.3.2 Opticals Properties

3.3.2.1 Optical Analysis by Ellipsometry

Figure 3.23 shows a good agreement between the experimental and theoretical spectra obtained for a Cr doped TiO$_2$ sample at 7 %. Good fit quality indicates that the Tauc–Lorentz model [7] properly describe the optical properties of TiO$_2$ films.

It is noted a strong variation of the refractive index in terms of Cr doping concentration (Fig. 3.24a).

This suggests that the increase of the refractive index is due to the densification of Cr–TiO$_2$ films when Cr concentration increases. Refractive index presents a

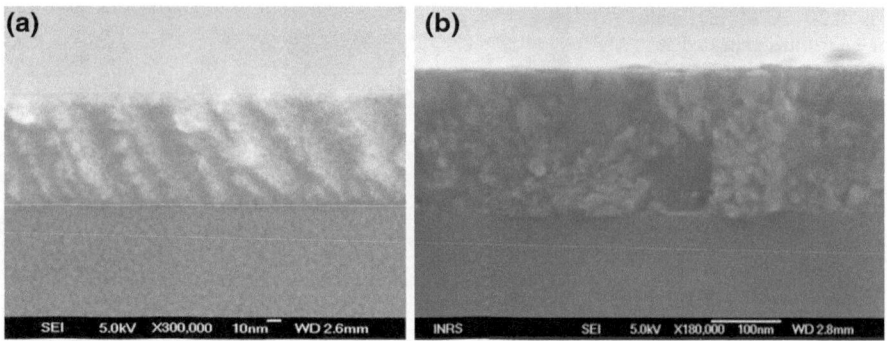

Fig. 3.22 Typical cross-section SEM micrographs of TiO$_2$ films (**a** and **b**) as annealed at 550 °C.
a Pur TiO$_2$, **b** TiO$_2$–Cr (17 at.%)

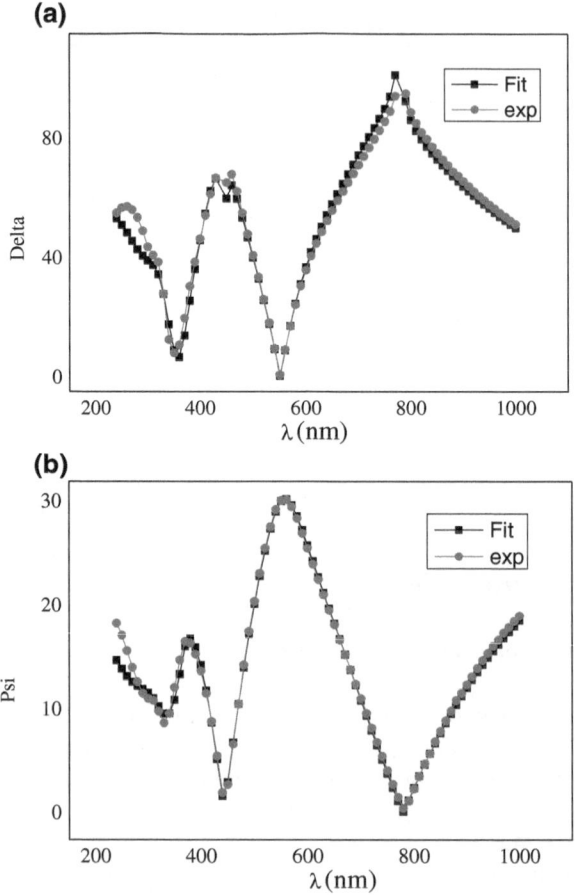

Fig. 3.23 Typical ellipsometry spectra and best fit of TiO$_2$ doped Cr (17 at.%) **a** Delta and **b** Psi

Fig. 3.24 Optical
properties and thickness
of TiO₂ film as a function
of Cr concentration. **a** The
refractive index and density
of films, **b** band gap and
c thickness

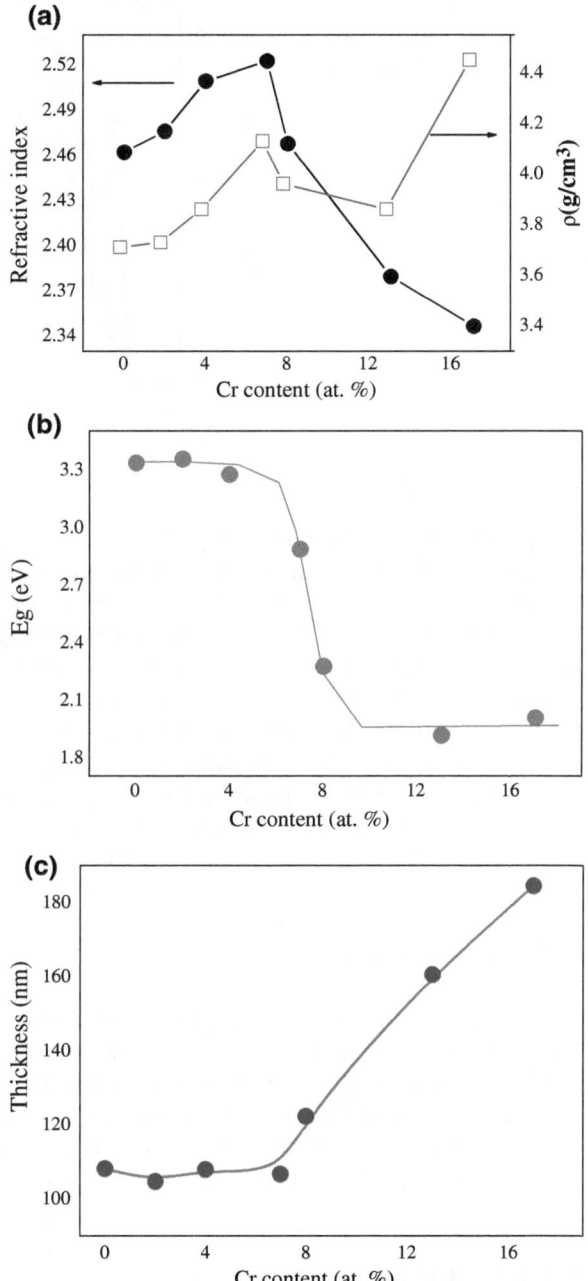

maximum (n = 2.52) for a Cr concentration of 7 % (concentration at which the
anatase-rutile transition occurs) and then decreases to reach 2.34 for a Cr concentration of 17 %. The decrease in the refractive index of Cr doped TiO₂ film (17 %)
can be explained by the oxidation of the Cr nanoparticles leading to a mixed phase

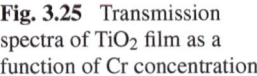
Fig. 3.25 Transmission
spectra of TiO$_2$ film as a
function of Cr concentration

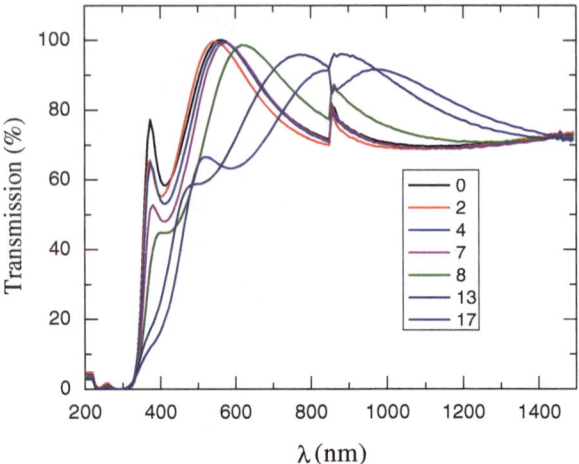

of type TiO$_2$–Cr$_2$O$_3$. Moreover, Fig. 3.24b shows the evolution of the TiO$_2$ films bandgap Eg with Cr content. The gap energy value varies between two values (3.3 eV for low Cr concentrations and 2 eV for higher concentrations), with an abrupt transition around 7 %. This evolution of the gap is owing to the presence of acceptor states in the bandgap. Transition metals could also make significant changes on the electronic structure of a crystalline material and thus on the value of the gap [17, 18]. The decrease in the gap makes TiO$_2$–Cr films more absorbent on a wide range of UV-Visible, constituting a major advantage in terms of photovoltaic applications. Figure 3.24c gives the variation of thickness with the incorporation of the metallic Cr in the TiO$_2$ films. We can notice that the estimated thicknesses are between 108 nm for pure TiO$_2$ and 185 nm for TiO$_2$–Cr (17 %).

3.3.2.2 Transmission

Figure 3.25 shows the effect of the Cr concentration on the optical transmission of TiO$_2$ films annealed at 550 °C. The films are quite transparent in the visible region for the pure TiO$_2$. Beyond 7 % Cr concentration, there was a decrease in the transmission of Cr-doped TiO$_2$ films (example: 17 %). This behavior is a consequence of the presence of Cr atoms in the lattice which not only lead to a lower transmission value, but also to a better Cr absorption for $\lambda < 600$ nm.

3.4 Conclusion

The titanium dioxide thin films in the TiO$_2$–Cr/intrinsic Si, TiO$_2$–Cr/SiP and TiO$_2$–Cr/quartz structures, were deposited by sputtering process. Sputtering plasma consists of a mixture of argon and oxygen. The pressure was maintained at 1.2 mTorr while the proportions of oxygen and argon in the mixture are 20 and 80 %, respectively. To determine the structural, optical and electronic characteristics of titanium dioxide thin films (TiO$_2$), some techniques such as: the DRX in grazing incidence, X-ray reflectometry, atomic force microscopy, Raman spectroscopy, ellipsomety, XPS and optical transmission and reflectivity measurements have been used.

The main results obtained during the analysis of titanium dioxide thin films doped with chromium can be summarized as follows:

- Titanium dioxide films crystallize at an annealing temperature of 550 °C; They consist of two phases: the anatase and rutile.
- Anatase-rutile phase change takes place for a Cr doping concentration of 7 % and an annealing temperature of of 550 °C; the rutile phase is the most stable phase, to achieve it you have to go at annealing temperatures higher than 700 °C.
- The metallic chromium increases with the increase of the doping concentration.
- The titanium dioxide thin films are transparent in the visible.
- Optical ellipsometry analysis allowed determination of the doping effect on optical properties. We study the effect of doping on the refractive index, extinction coefficient and the gap.

In the next chapter we will use these results to study and explain the principle of operation of the Cr-doped TiO$_2$ sensor and its response under a reducing gas such as ethanol. We will also explore the effects of this material on the photoluminescence of porous silicon and its possibly application in solar cells.

References

1. Hafidi K et al (2004) Déposition par pulvérisation cathodique radio fréquence et caractérisation électronique, structurale et optique de couches minces du dioxyde de titane. Act Passive Electron Compon 27:169
2. Crose P, Nevot L, Padro B (1972) Contribution a l'étude des couches minces par réflexion spéculaire de rayons X. Rev. d'Optique Appliquée 3:37
3. Djaoued Y, Badilescu S, Ashrit PV, Bersani D, Lottici PP, Robichaud JJ (2002) Study of anatase to rutile phase transition in nanocrystalline titania films. Sol-Gel Sci Technol 24:255
4. Jellison GE Jr (1993) Data analysis for spectroscopic ellipsometry. Thin Solid Films 234:416
5. Jellison GE Jr, Modine FA (1996) Parameterization of the optical functions of amorphous materials in the interband region. Appl Phys Lett 69:371
6. Jellison GE Jr, Merkulov VI, Puretzky AA, Geohegan DB, Eres G, Lowndes DH, Caughman JB (2000) Characterization of thin-film amorphous semiconductors using spectroscopic ellipsometry. Thin Solid Films 68:377
7. Hajjaji A, Labidi A, Gaidi M, Ben Rabha M, Smirani R, Bejaoui A, Bessais B, El Khakani MA (2011) Structural, optical and sensing properties of Cr-doped TiO2 thin films. J Sens Lett 9:1697
8. Ginier A (1964) Théorie et technique de la radiocristallographie. Dunod Paris, 462
9. Bessaïs B (1992) Elaboration par sérigraphie et caractérisation électrique et optique de couches minces d'ITO (In$_2$ O$_3$ (Sn)). Thèse doctorat, Faculté de Sciences de Tunis

10. Carp O, Huisman CL, Reller A (2004) Photoinduced reactivity of titanium dioxide. Prog Solid State Chem 32:33
11. Arrayo R, Cordoba G, Padilla J, Lara VH (2002) Influence of manganese ions on the anatase–rutile phase transition of TiO2 prepared by the sol–gel process. Mater Lett 54:397
12. Yuan X-L, Zhang J-L, Anpo M, He D-N (2010) Synthesis of Fe^{3+} doped ordered mesoporous TiO_2 with enhanced visible light photocatalytic activity and highly crystallized anatase wall. Res Chem Intermed 36:83
13. Kim EJ et al (2001) Microstrucral changes of microemulsion-mediated TiO_2 particles during calcinations. Mater Lett 49:244
14. Catherine P (2006) Syntheses de nanocristaux de TiO_2 anatase a distribution de taille controlée. Influence de la taille des cristallites sur le spectre Raman et étude des propriétés de surface. Thèse doctorat, Université de Bourgogne, Paris
15. Hajjaji A, Gaidi M, Bessais B, El Khakani MA (2011) Effect of Cr incorporation on the structural and optoelectronic properties of TiO_2: Cr deposited by means of a magnetron co-sputtering process. J Appl Surf Sci 257:10351
16. Luu CL, Nguyen QT, Ho ST (2010) Synthesis and characterization of Fe-doped TiO2 photocatalyst by the sol-gel method. Adv Nat Sci: Nanosci Nanotechnol 1:015008
17. Singh D, Singh N, Sharma SD, Kant C, Sharma CP, Pandey RR, Saini KK (2011) Bandgap modification of TiO_2 sol-gel films by Fe and Ni doping. J Sol-Gel Sci Technol 58:269
18. Anpo M (2000) Use of visible light. Second-generation titanium oxide photocatalysts prepared by the application of an advanced metal ion-implantation method. Pure Appl Chem 72:1787

Chapter 4
Gas Sensors and Photo-Conversion Applications

Abstract First, this chapter investigates the electrical behavior as well as stability and reproducibility of Cr doped TiO_2 thin films against ethanol gas. Second the TiO_2-Cr/SP/Si structures are studied using different techniques such as AFM, SEM, photoluminescence, lifetime and laser-induced current (LBIC). Finally, the optoelectronic and photosensibility properties of such structures based on Cr doped TiO_2 revealed potential applications in photovoltaic solar cells.

Keywords Conductance · Gas sensors · Ethanol · Sensitivity · Microstructure · Atomic force microscopy (AFM) · Roughness parameter · Light Beam Induced Current (LBIC) measurements · Reflectivity · Photoluminescence · Photo-conversion

4.1 Introduction

In this chapter, we present, in a first part, the electrical behaviour of the TiO_2 doped films for different Cr concentration in the presence of ethanol gas. In the second part, we will be interested to the deposition of TiO_2–Cr films on monocrystalline and multicrystalline porous Silicon. Also, we will study the influence of Cr doping on the characteristics of the TiO_2–Cr/PS/Si type components, using different techniques such as AFM, SEM, photoluminescence, lifetime and laser induced current (LBIC). Indeed, in the first part, the objective is to test and study the sensitivity, stability and reproducibility of electrical measurements on Cr doped TiO_2 films in the presence of ethanol; the effect of the Cr concentration on sensitivity, response and recovery time of TiO_2 based thin-film sensor are then analyzed. In the second part, we will be interested in the effect of the Cr incorporation on the optoelectronic properties of the TiO_2/PS/Si (mono- or multi-crystalline) structure.

© The Author(s) 2015
A. Hajjaji et al., *Chromium Doped TiO2 Sputtered Thin Films*,
SpringerBriefs in Manufacturing and Surface Engineering,
DOI 10.1007/978-3-319-13353-9_4

4.2 Detection Properties of Cr Doped TiO$_2$ Films for Ethanol

4.2.1 Tests Under Ethanol

Detection properties of samples for ethanol were studied using the experimental setup presented in Fig. 2.7 (Chap. 2). The two mass flow controllers (D1, D2) are used to adjust the dry air flow in the range 10–94 L h^{-1}. In the measurement chamber, air plays the role of carrier gas for ethanol vapors that come from a bottle placed in a thermostated bath with temperature set at 30 °C, in order to adjust the partial pressure of ethanol vapour. In these conditions, from the Eq. 4.1 is assumed the different ethanol concentrations carried by dry air [1, 2],

$$[C]\% = \left[\frac{\mu \cdot d_1}{\mu \cdot d_1 + d_1 + d_2}\right] \cdot 100 \tag{4.1}$$

where μ is the molar fraction of ethanol vapour with a given dilution in the bath by:

$$\mu = \frac{P_{vap}}{P_{atm}} \tag{4.2}$$

P_{atm} is the atmospheric pressure and P_{vap} is the partial vapor pressure at temperature T_{vap}. Partial pressures of ethanol vapors that we used to test sensors responses are deducted directly from the curve presented in Fig. 4.1.

For a fixed temperature $T_{vap} = 30$ °C, the variation of the concentration depends only on the values of flows d_1 and d_2 (dry air from flowmeter d_1 is loaded by vapor passing in a bath of pure organic liquid). The latter is maintained at a fixed evaporation temperature in order to adjust the vapor pressure. This vapor mixes with the second

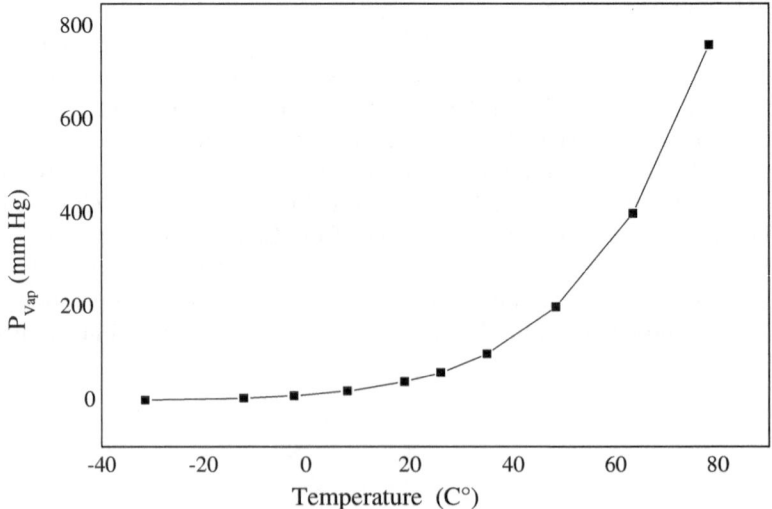

Fig. 4.1 Change in ethanol vapor pressure with temperature

dry air flow coming from second flowmeter d$_2$. The air flow is kept constant whatever the used gas. The working temperature of the sample was maintained at T = 200 °C, by applying a voltage of 1 V on the heating element in the test chamber. Low sensitivities were obtained for doped and undoped TiO$_2$ layers with 0.1 % ethanol mixture.

4.2.2 Electrical Characterizations Under Ethanol

In Chap. 3, we detailed the microstructural and optical characteristics of the Cr-doped TiO$_2$ films deposited at ambient room temperature. We have shown that the latter have a semi-amorphe crystallographic structure and the increase of doping (from 0 to 17 %) decreases the band gap energy value from 3.31 to 1.89 eV. The level of Cr doping controls the electrical conduction of TiO$_2$ films. A change of type (N type to P type) was observed for the two Cr concentrations 13 and 17 % atomic. This behavior change is interesting in the detection of reducing or oxidizing gases.

4.2.3 Conductance Measurement

Figure 4.2 shows the response of a pure TiO$_2$ sensor for an ethanol concentration of 0.1 % at a working temperature of 200 °C. It should be noted that this response for ethanol are quite reproducible over a period of approximately 60 min. One note that the signal drop should take a longer time.

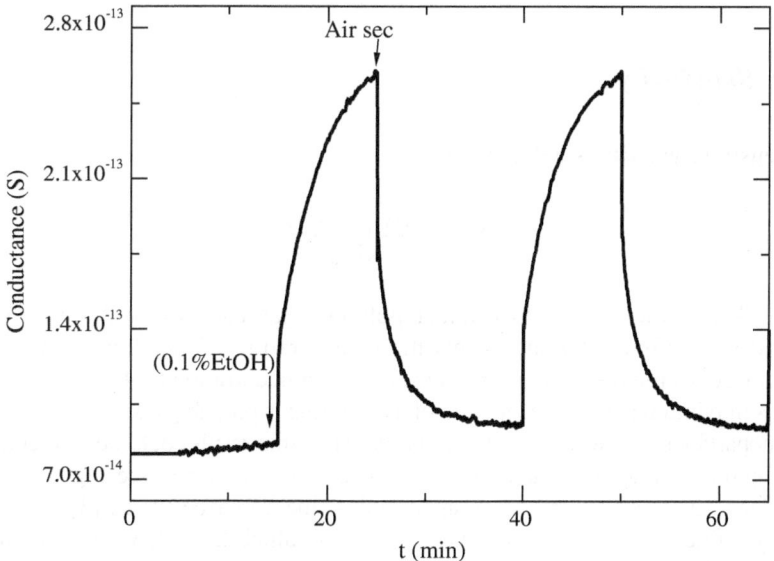

Fig. 4.2 Dynamic measurement of the electrical conductance of pure TiO$_2$ under ethanol

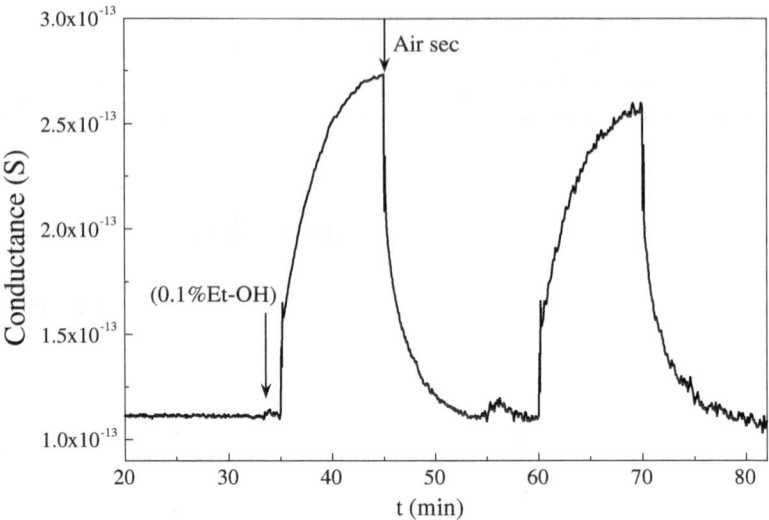

Fig. 4.3 Dynamic measurement of the electrical conductance of TiO$_2$ doped Cr (8 at.%) of ethanol

In the same way, Fig. 4.3 shows that the incorporation of transition metal such as Cr (8 at.%) strongly changes the sensor response in the presence of ET-OH gas compared to the pure TiO$_2$. We have also noted a return to base line for the doped TiO$_2$ films. This can be explained by the decrease in grain sizes, which induce a larger specific surface able to adsorb more oxygen during the recovery time. However, one may note the non-reproducibility of conductivity after a repetition time of about 25 min.

4.2.4 Sensitivity

The sensor sensitivity is defined by:

$$S_{Vap} = \frac{G_{Vap} - G_{air}}{G_{Vap}} \qquad (4.3)$$

where G$_{vap}$ is the conductance under pollutant gas and G$_{air}$ is the conductance under dry air. Figure 4.4 shows that the incorporation of metallic Cr in the TiO$_2$ films improves the response of the sensor at Cr concentration of 4 %.

This maximum at Cr concentration of 4 % atomic is probably due to the presence of Cr nanoparticles that would reach an optimal size, this result could be also correlated to the grain morphology that varies with the Cr concentrations (see Fig. 3.20, Chap. 3). Indeed, there is appearance of small grains (the Yamazoe works [3] have shown that crystallite sizes <10 nm, samples of polycrystalline SnO$_2$ have best sensitivities with reducing gases), to the surface adsorption increases, and then response to ethanol vapor increases, involving a maximum of sensitivity for a Cr concentration of 4 %.

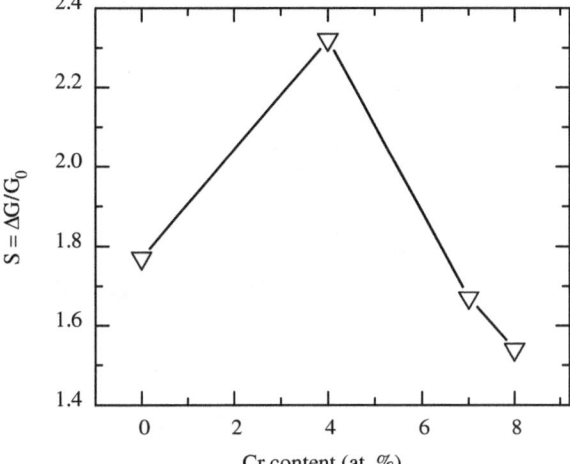

Fig. 4.4 Sensor response of TiO₂ thin films as a function of Cr concentration

The same behavior was reported during the incorporation of nanoparticles (Au and Pt) in the SnO₂ films used in the detection of CO gas [4–7]. At this optimal concentration, the particle sizes are about 1–2 nm. Figure 4.5 shows that for 4 % atomic, the concentration of Cr_2O_3 oxide reached a maximum; the same phenomenon was observed by Gaidi et al. on SnO₂ films doped with Platinum and palladium where the

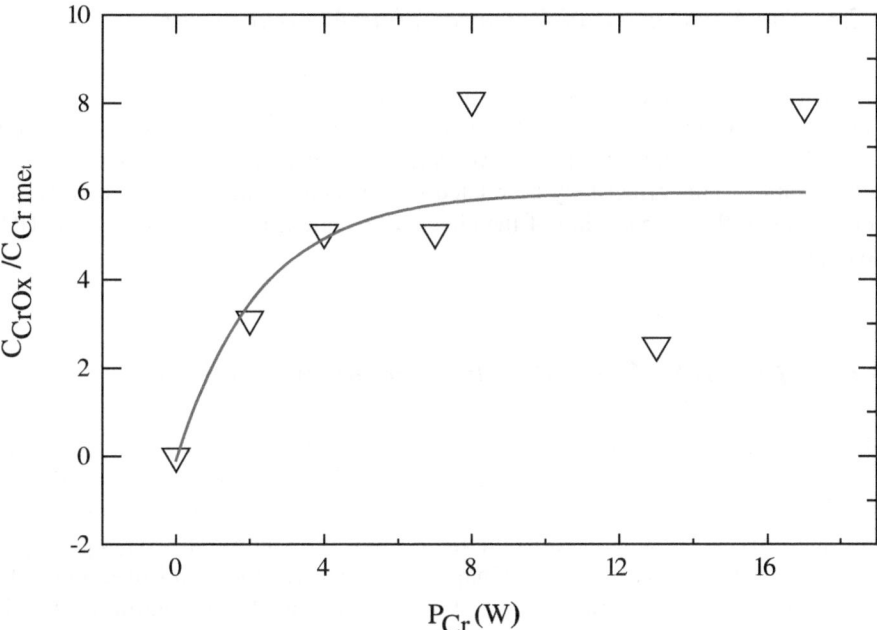

Fig. 4.5 Concentration of Cr in oxidation state (CrOₓ) compared to its concentration in metallic (Crₘₑₜ)

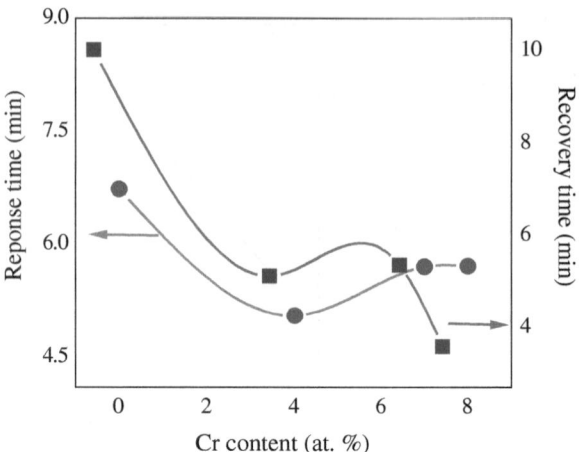

Fig. 4.6 Sensing properties of TiO$_2$ thin films as a function of Cr concentration response and recovery time

effect of doping on the response of the sensor under gas is more pronounced when the additive metallic is in oxidation state [6, 7].

4.2.5 The Response and Recovery Time Measurement

Figure 4.6 shows the response time and the recovery time in terms of Cr content. These times are measured when the conductance achieved 90 % of its maximum or 90 % of its minimum. These two parameters decreased with increasing Cr concentration. This effect may be related to the catalytic activity of metal, which contributes to the acceleration of the chemical reaction on the surface of the TiO$_2$ layer [8].

4.2.6 Effect of Cr Content on the Conductance Behavior

Figure 4.7a, b show the variation of conductance for two Cr concentrations (13 and 17 %). There is a reverse behavior compared to what was observed for (Cr 4 %). Indeed, the introduction of ethanol gas induces a decrease in conductance and this would be linked to the increase of the content in CrO$_2$ and Cr$_2$O$_3$ (based on XPS measurements of Chap. 3). In fact, for this level of doping, the layers change from N doping type to P character type. The evolution reverse in the conductance is probably due to the presence of these two phases acting as acceptors [9].

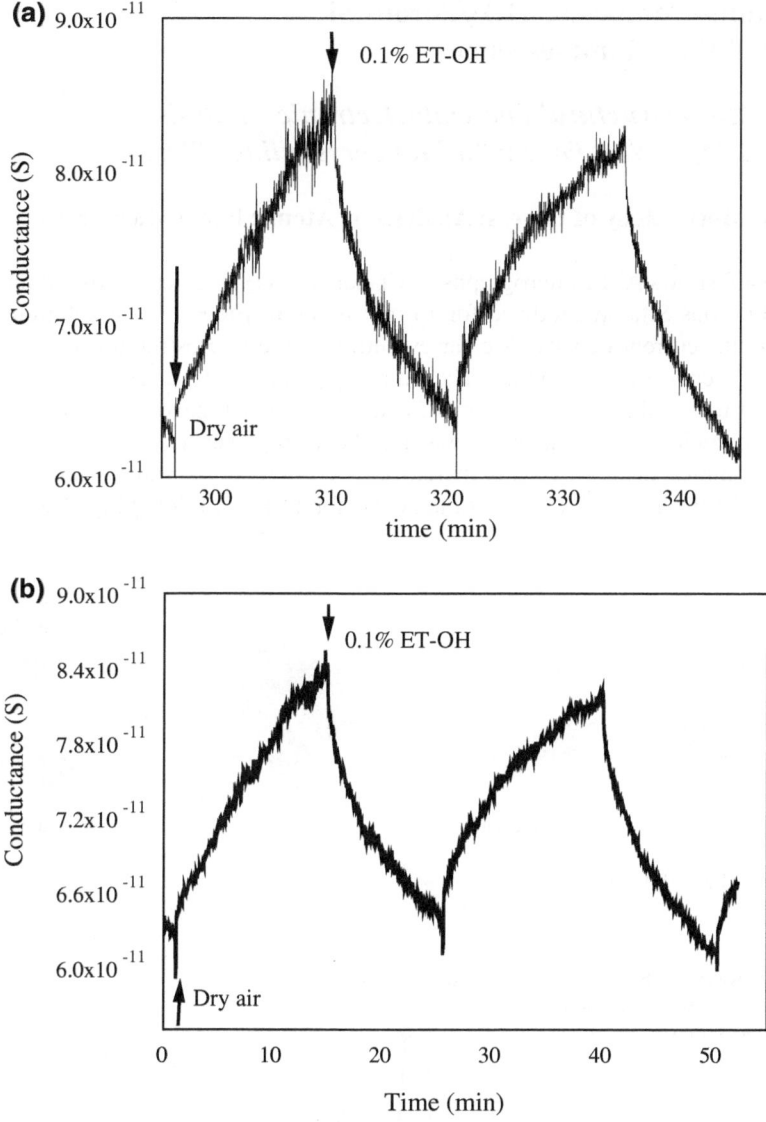

Fig. 4.7 Dynamic measurement of the electrical conductance under ethanol of TiO₂ doped. **a** Cr (13 at.%), and **b** Cr (17 at.%)

4.3 Nano-Composite TiO$_2$/Porous Si and Photoluminescence

4.3.1 Microstructural and Optoelectronic Analysis TiO$_2$–Cr/Si–Porous/Si Monocrystalline Films

4.3.1.1 Morphology of Layers: Analysis by Atomic Force Microscopy

Figure 4.8 shows AFM micrographs of Cr-doped TiO$_2$ samples deposited on single crystal porous Silicon anodized for 19 min in a (HF:EtOH) = 1:4 solution with a 15 mA/cm^2 current density. A clear evolution of the microstructure was observed with the increase of Cr doping. The pure TiO$_2$ film shows a less porous and less grainy structure than a doped films (2 and 17 %), which have rough surfaces with a large particles. This analysis shows that the average surface roughness of TiO$_2$–Cr/PS samples decreases with increasing Cr content; its value decreases from 100 nm for the TiO$_2$–Cr (2 %) film to 30 nm for TiO$_2$–Cr/(17 %) Fig. 4.9.

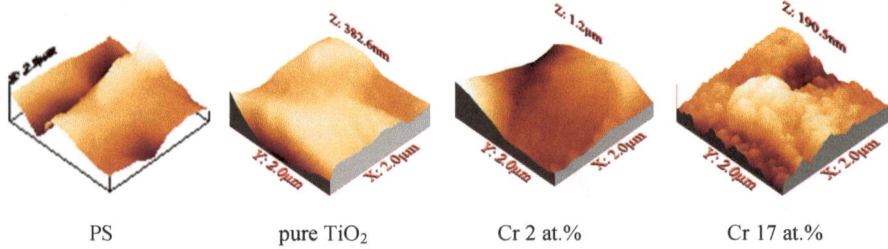

PS   pure TiO$_2$ Cr 2 at.%   Cr 17 at.%

Fig. 4.8 AFM (3D) scans

Fig. 4.9 RMS of PS/TiO$_2$: Cr/c–Si samples as a function of Cr concentration (at.%)

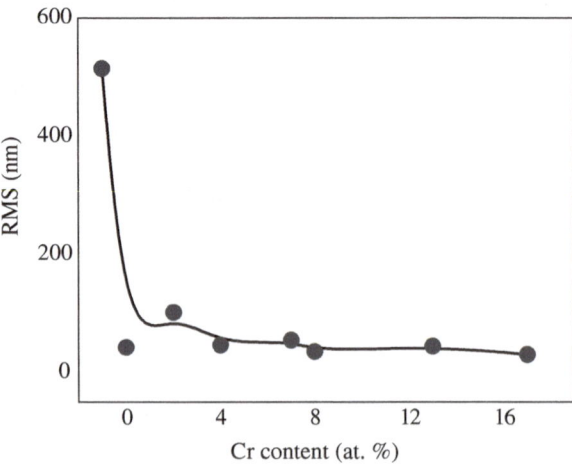

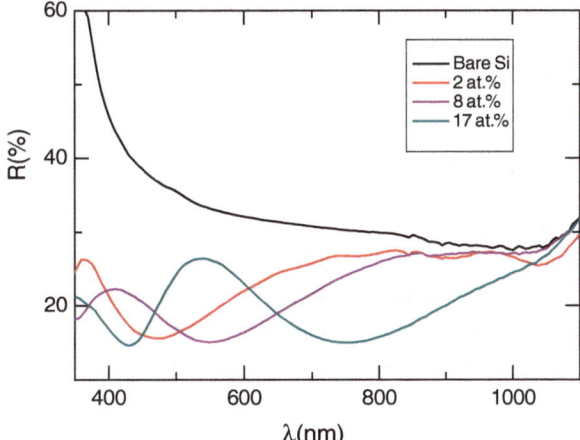

Fig. 4.10 Evolution of the total reflectivity as a function of Cr concentration (at.%)

4.3.1.2 Reflectivity

Figure 4.10 shows the reflectivity spectra of undoped and Cr-doped TiO_2 layers and deposited on a monocrystalline PS substrate. These spectra show that TiO_2 films can play the role of antireflective layers. Indeed, in the wavelength range between 350 and 700 nm reflectivity decreases from 41 % for c–Si to nearly 17 % for Cr doped films (17 %). This decrease of the reflectivity is probably due to the formation of a rough structure suitable for light trapping.

4.3.1.3 Photoluminescence

Figure 4.11 shows photoluminescence (PL) spectra of Cr doped TiO_2 films deposited on a PS/monocrystalline Si substrate and annealed at temperature of 550 °C. It is found

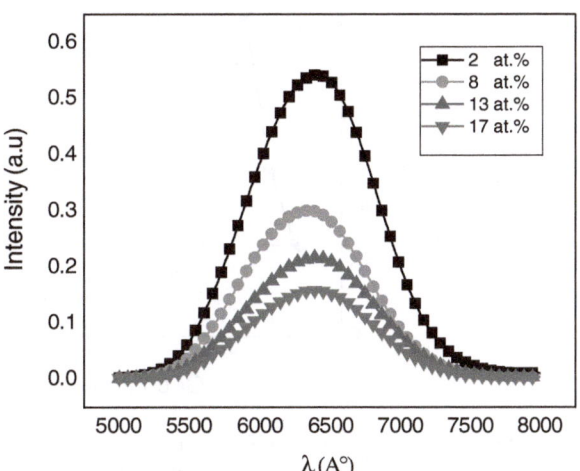

Fig. 4.11 PL spectra of PS/TiO₂:Cr treated samples as a function of Cr content

that the most PL intensity corresponds to a Cr concentration of 2 %. PL spectra show a maximum between 550 and 700 nm, depending on the Cr concentration. Beyond a concentration of 2 %, the PL intensity decreases with the doping concentration.

4.3.1.4 LBIC Measurements

Light Beam Induced Current (LBIC) measurements have been obtained using the wavelength of a He–Ne laser 632 nm. The effective diffusion length of minority carriers (L) is one of the photovoltaic parameters used to quantify the quality of the material. The effective diffusion length L is calculated from the LBIC data by using the following expression [10, 11]:

$$I_{LBIC}(x) = A \exp(-\frac{x}{L})x^{-n} \qquad (4.4)$$

where x is the distance between the aluminum electrodes [12] and the laser beam, A is a constant depending mainly on the intensity of the laser beam and the penetration depth [11] and n is a constant function of the surface recombination velocity.

Figure 4.12 shows the evolution of the diffusion length of the monocrystalline silicon with Cr concentration. It is found that the greater diffusion length is associated to a low value of Cr concentration (2 %); This tendency is in concordance with the PL measurements (Fig. 4.11). The increase in Cr concentration decreases the diffusion length, this can be explained by the strong Cr doping that induces defects. These defects come from various Cr oxides formed during the preparation of doped films [13].

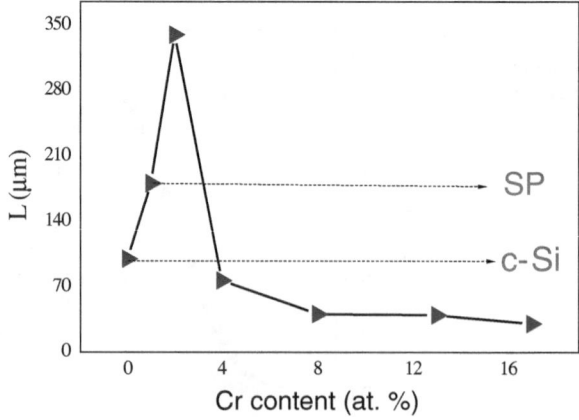

Fig. 4.12 c–Si minority carrier diffusion length (L) as a function of Cr concentration (at.%). L for untreated c–Si and PS silicon treated sample is given for comparison

4.3.2 *Microstructural and Optoelectronics Analysis of the TiO$_2$–Cr/Porous Multicrystalline Si Films*

4.3.2.1 Layers Morphology: Analysis by Atomic Force Microscopy

Figure 4.13 shows the surface microstructure of the TiO$_2$–Cr/PS/multicrystalline Si depending on the concentration of metallic Cr. The surface of pure TiO$_2$ film is porous, however the structure of doped films becomes grainy from Cr doping equal to 7 %.

Figure 4.14 shows the evolution of the TiO$_2$ films roughness with Cr doping.

Similarly, we note that the roughness increases for a Cr concentration of 2 %, beyond this value the roughness decreases considerably with Cr content.

4.3.2.2 Layers Morphology: Analysis by Scanning Electron Microscope

SEM images (Fig. 4.15) show that TiO$_2$ film deposited on PS/Si–mc has an ordered rough surface that could be adapted for light trapping.

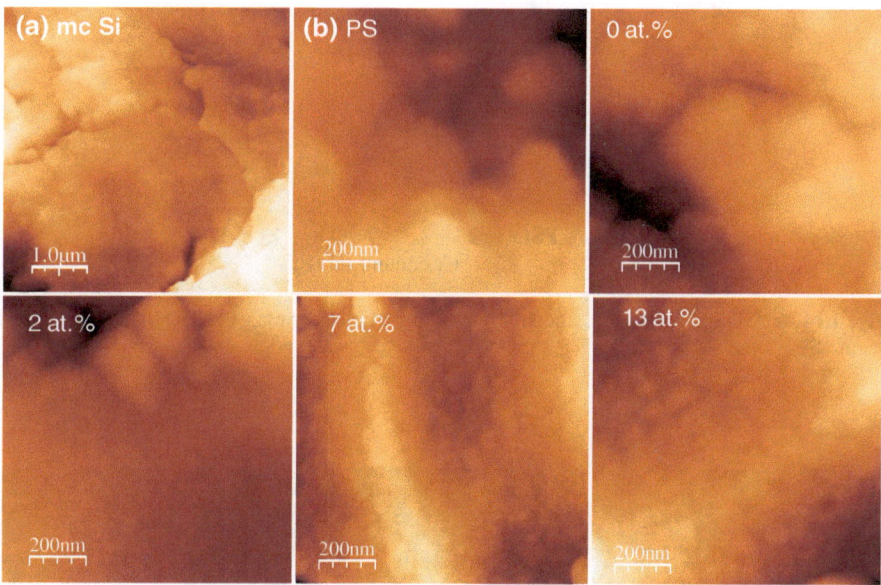

Fig. 4.13 AFM images of TiO$_2$–Cr/PS/mc–Si film

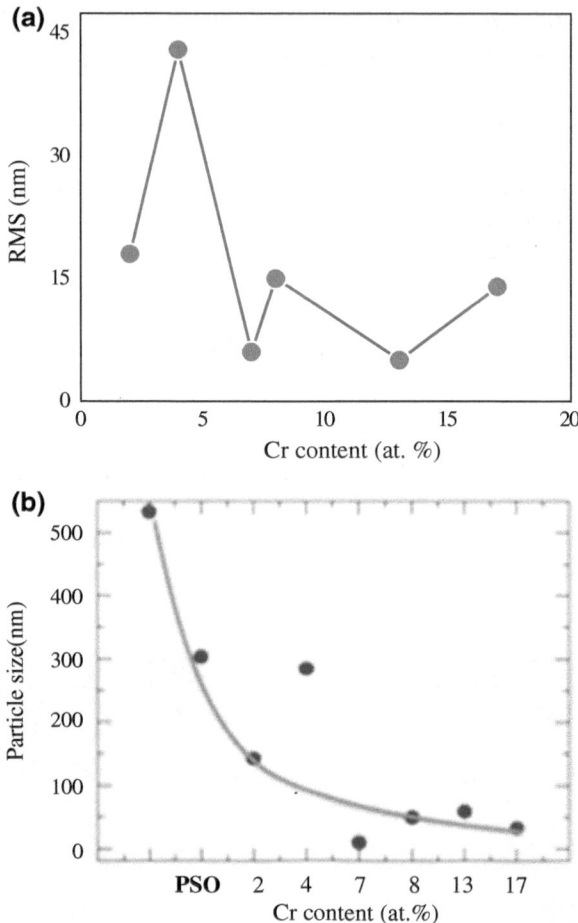

Fig. 4.14 **a** Average roughness, and **b** size of the particles forming the TiO_2–Cr deposited on PS/mc-Si

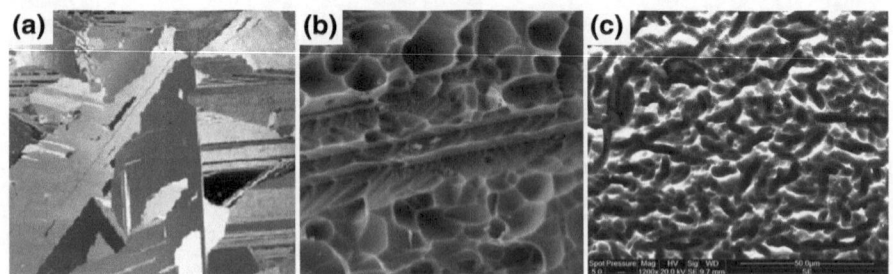

Fig. 4.15 SEM morphology of **a** mc-Si, **b** PS/mc-Si and **c** TiO_2–Cr/PS/mc-Si

Fig. 4.16 Total reflectivity spectra as a function of Cr content

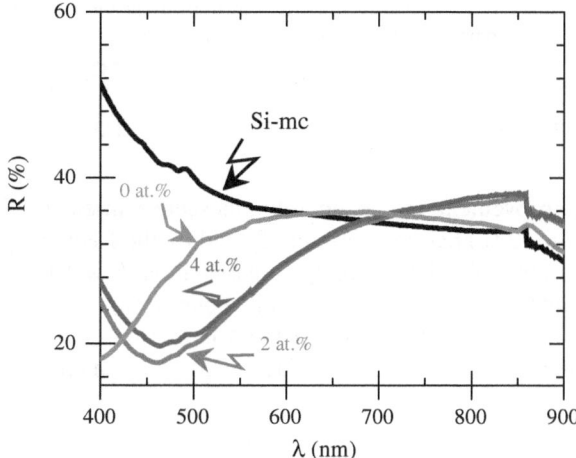

4.3.2.3 Reflectivity

Figure 4.16 shows the reflectivity spectra of pure and Cr doped TiO$_2$ layers deposited on PS substrate from multicrystalline Si. These spectra show that TiO$_2$ films play the role of antireflective layers. For wavelength range between 350 and 700 nm, the reflectivity decreases to 41 % for the Si–mc to nearly 17 % for the Cr doped films (17 %). This decrease of reflectivity is attributed to the formation of a rough structure (Fig. 4.15).

4.3.2.4 Lifetime

The effective lifetime τ_{eff} is given by [14]:

$$\frac{1}{\tau_{eff}} = \frac{1}{\tau_{bulk}} + \frac{2S}{W} \tag{4.5}$$

where S is the recombination velocity and W is the thickness of the Si–mc substrate. The evolution of the minority carriers effective lifetime of Si–mc after treatment with the PS, pur TiO$_2$/PS and TiO$_2$–Cr (2 %)/PS is summarized in Table 4.1.

The results indicate an improvement of this parameter compared to the value obtained from the untreated multcristalline Si. In fact, it increases from 23 μs for the mc–Si layer to an optimum value of 733 ms for the TiO$_2$–Cr film (2 %). After this value, the lifetime decreased again with the increase of the Cr concentration. This improvement may be attributed to the surface passivation and the reduction of the recombination velocity due to the treatment of the porous Silicon layer and the TiO$_2$–Cr/PS film as a dual layer [15]. This phenomenon has been reported [8] by Vasquez et al. [16] which indicated that porous Si oxidation leads to a strong

Table 4.1 Evolution of the effective lifetime of the minority carriers of mc–Si after treatment with PS, pure TiO_2/PS and TiO_2–Cr (2 at.%)/PS

	mc–Si	PS/mc–Si	TiO_2/PS/mc–Si	TiO_2:Cr (2 at.%)/PS/mc–Si
Life time (µs)	2	23	233	733

improvement of the photoluminescence intensity. The lifetime for the TiO_2–Cr film (2 %) has been also confirmed by the LBIC current measurements at the level of the grain. The current decreases about 5×10^{-8} A for the non-doped TiO_2 film and 9×10^{-8} A for the Cr doped one (2 %).

Figure 4.17 shows that there is a strong correlation between the PL intensity and the Cr concentration. The highest PL intensity is obtained for the TiO_2 film

Fig. 4.17 **a** PL spectra and **b** correlation between the lifetime and intensity of TiO_2 doped Cr/PS–mc-Si samples

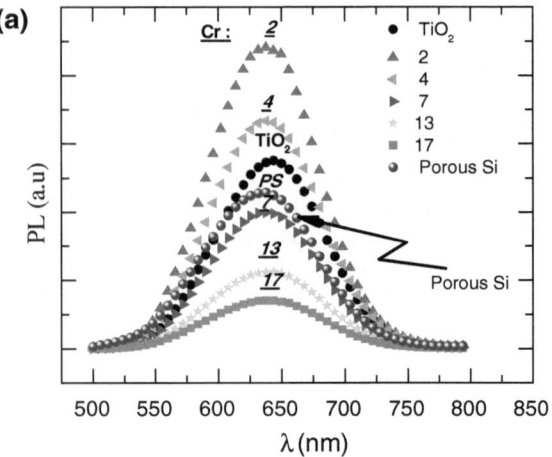

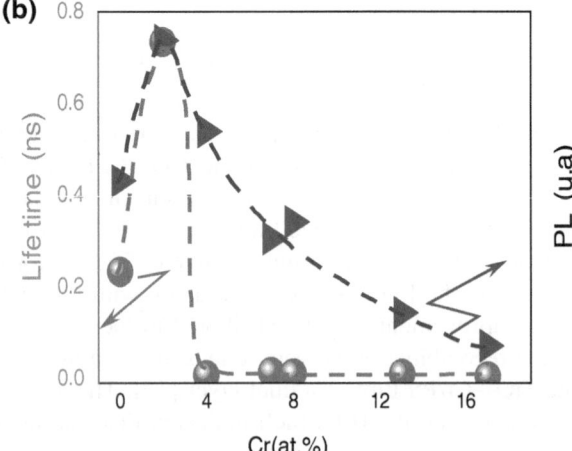

Fig. 4.18 Experimental LBIC spectra: **a** experimental spectra, **b** fitting the current LBIC in the case of pure TiO$_2$/PS/mc-Si, and **c** TiO$_2$ doped Cr (2 at.%)/PS/mc-Si

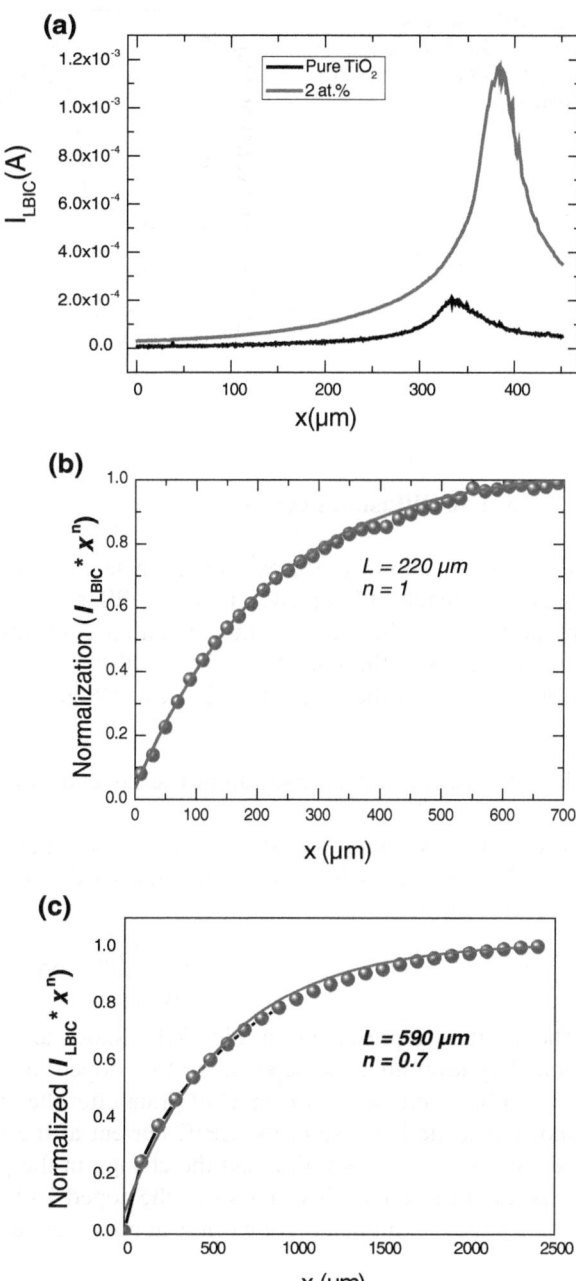

doped with Cr (2 %). Although the position of the PL line is almost independent of the Cr content, the improvement of the PL response is due to the passivation of the PS by the TiO$_2$ layers doped with Cr. One note also a significant correlation between the lifetime and the PL intensity (Fig. 4.17b) [17].

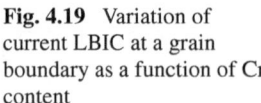

Fig. 4.19 Variation of current LBIC at a grain boundary as a function of Cr content

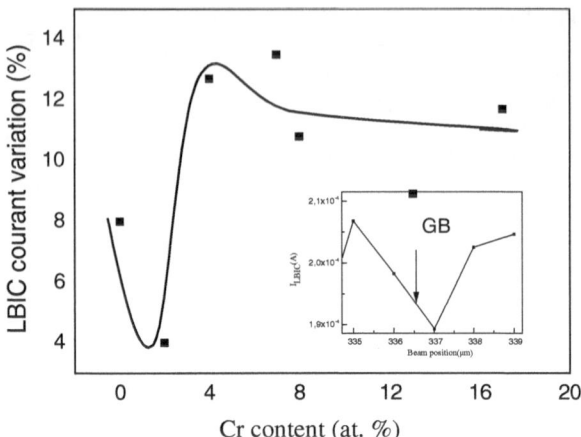

4.3.2.5 The Diffusion Length

The analysis shows a decrease of the n factor from 1 for the pure TiO_2/PS–mc to approximately 0.7 for TiO_2–Cr (2 %)/PS–mc. As the recombination velocity is proportional to the factor n, this evolution confirms results from analysis of the minority carriers lifetime. The effective diffusion length (Fig. 4.18) was improved from 220 μm for the pure TiO_2/PS–mc to 590 μm for TiO_2–Cr (2 %)/mc–PS.

4.3.2.6 Local Characterisation at the Level of the Grains Joint

In Fig. 4.19, we represent the variation of the measured current at grain boundary of the Si-mc in the case of pure and Cr doped TiO_2 deposited on the porous multicrystalline silicon:

$$\frac{I_{max} - I_{min}}{I_{max}} * 100 \qquad (4.6)$$

The profiles represented in Fig. 4.19 show an increase of LBIC current at boundary level after the deposition of Cr doped titanium dioxide films. The LBIC current has increased to the level of grain after the incorporation of Cr metal. This shows that the increase in the LBIC current at the level of the grain boundary is due to a volume passivation and the change in the physical and chemical characteristics of the latter. This shows that the doped TiO_2 films passivate in volume and reduces the recombination of carriers at boundary level.

4.4 Conclusion

In the first part of this chapter we showed that the response of TiO_2 based sensor to ethanol vapors dependent on Cr concentration in the TiO_2 films. We found that the sensors are fairly stable, and the TiO_2 film doped with Cr (4 %) presents the best response for ethanol. Furthermore, for 0.1 % ethanol, a maximum sensitivity is reached for an operating temperature of 200 °C.

In the second part, we have shown that the titanium dioxide obtained by sputtering and deposited on monocrystalline porous Silicon substrates can effectively contribute to the reduction of reflection losses. The deposition of the Cr doped TiO_2 film on the PS improves the diffusion length and lifetime of the silicon substrate.

As conclusion, this analysis shows that the Cr-doped TiO_2 films combined with porous Silicon enhances the optoelectronic properties of mono and multicrystalline Si. A dramatic increase in the multicrystalline Si lifetime equal to 733 µs was reached. The experimental results suggest that the TiO_2–Cr/PS induces a good multicrystalline and monocrystalline silicon surface passivation.

References

1. Abdelghani A, Jaffrezic-Renault N (2001) SPR optical fibre sensor sensitized by fluorosiloxane polymers. Sens Actuators B 74:117
2. Labidi A, Bejaoui A, Ouali H, Akkari FC, Hajjaji A, Gaidi M, Kanzari M, Bessais B, Maaref M (2011) Dry air effects on the copper oxides sensitive layers formation for ethanol vapor detection. Appl Surf Sci 257:9941
3. Yamazoe N (1991) New approaches for improving semiconductor gas sensors. Sens Actuators B 5:7
4. Dolbec R, El Khakani MA (2005) Pulsed laser deposited platinum and gold nanoparticles as catalysts for enhancingthe CO sensitivity of nanostructured SnO_2 sensors. Sens Lett 3:216
5. Dolbec R, El Khakani MA (2007) Sub-ppm sensitivity towards carbon monoxide by means of pulsed laser deposited SnO_2: Pt based sensor. Appl Phys Lett 90:173114
6. Gaidi M, Labeau M, Chenevier B, Hazemann JL (1998) In-situ EXAFS analysis of the local environment of Pt particles incorporated in thin films of SnO_2 semiconductor
7. Gaidi M, Chenevier B, Labeau M (2000) Electrical properties evolution under reducing gaseous mixtures (H_2, H_2S, CO) of SnO_2 thin films doped with Pd/Pt aggregates and used as polluting gas sensors. Sens Actuators B 62:43
8. Hajjaji A, Labidi A, Gaidi M, Ben Rabha M, Smirani R, Bejaoui R, Bessais B, El Khakani MA (2011) Structural, optical and sensing properties of Cr-doped TiO_2 thin films. J Sens Lett 9:1697
9. Ruiz AM et al (2003) Cr-doped TiO_2 gas sensor for exhaust NO2 monitoring. Sens Actuators B 93:509
10. Davidson SM, Dimitriadis CA (1980) Advances in the electrical assessment of semiconductors using the scanning electron microscope. J Microsc 118:275
11. Ioannou DE, Gledhill RJ (1983) SEM-EBIC and traveling light spot diffusion length measurements—normally irradiated charge-collecting diode. IEEE Trans Electron Devices 30:577
12. Khedher N, Hajji M, Bouaicha M, Boujmil MF, Ezzaouia H, Bessais B (2002) Improvement of transport parameters in solar grade monocrystalline silicon by application of a sacrificial porous silicon layer. Solid State Commun 123:7
13. Hajjaji A, Ben Rabha M, Janene N, Dimassi W, Bessais B, El Khakani MA, Gaidi M (2012) Effect of dual treatment based on porous silicon and sputter-3 deposited TiO2 doped Cr film on the optoelectronic properties of monocrystalline Si. J Science JETC Marr 12:1
14. Sproul AB (1994) Dimensionless solution of the equation describing the effect of surface recombination on carrier decay in semiconductors. J Appl Phys 76:2851
15. Ben Rabha M, Bessaïs B (2010) Enhancement of photovoltaic properties of multicrystalline silicon solar cells by combination of buried metallic contacts and thin porous silicon. Sol Energy 84:486

16. Vasquez-A MA, Aguila Rodrigueza GA, Garcia-Salgado G, Romero-Paredesa G, Pena-Sierra R (2007) FTIR and photoluminescence studies of porous silicon layers oxidized in controlled water vapour conditions. Rev Mex Fis 53:431
17. Hajjaji A, Ben Rabha M, Janene N, Gaidi M, Bessais B, El Khakani MA (2012) Effect of dual treatment based on porous silicon and sputter-deposited TiO_2 doped Cr film on the opto-electronic properties of monocrystalline Si. Appl Surf Sci 258:8046

Chapter 5
TiO$_2$ Photocatalysis

Abstract This work is completed with a summary and a brief outlook for the further improvement of the photocatalytic performance when this chromium element is incorporated into TiO$_2$ oxide growing on quartz and porous multicrystalline silicon substrates. These photosensitivity tests emerged as a new way to show the important role of the structural modification and microstructure of such Cr doped TiO$_2$ sputtered thin films.

Keywords Cr-doped TiO$_2$ · Si-porous/Si and quartz substrates · Urbach energy · Photocatalysis · Absorbance · Amido Black · Microstructure

5.1 Introduction

The photocatalytic properties of semiconductor materials are closely related to its crystal quality, chemical composition and phase purity, it is essential to understand the detailed structural and optical properties before their further application as sensing device. This chapter will investigate the effect of Cr-doped TiO$_2$ sputtered thin films deposited on different substrates. It is well known that various products using photocatalytic functions have been recently used and commercialized. TiO$_2$ has indeed the most efficient photoactivity, the highest stability and also the lowest cost, Thus, this binary oxide is almost the only material suitable for industrial use at present and also probably in the near future owing to its safety to humans and the environment. Two photochemical reaction types proceeding on a TiO$_2$ surface when irradiated with ultraviolet irradiation: the first concerns the photo-induced redox reactions related to adsorbed substances, and the second is due to the photo-induced hydrophilic conversion of this oxide. The latter was found at the end of the century. Also, we have witnessed a revival and a rapid expansion in titanium oxide (TiO$_2$) based nanoscale as well as thin film structures deposited on various substrates and having high photocatalytic devices. The present work aims to present photocatalytic activity of Cr-doped TiO$_2$ thin films deposited on quarts as well as on multicrystalline porous silicon (PS) substrates. The later regarding PS substrate

© The Author(s) 2015

A. Hajjaji et al., *Chromium Doped TiO$_2$ Sputtered Thin Films*,
SpringerBriefs in Manufacturing and Surface Engineering,
DOI 10.1007/978-3-319-13353-9_5

has been widely explored for its strong visible room-temperature photoluminescence (PL) and for its high potential application in photovoltaic (PV) [1, 2]. TiO_2 is often used in screen-printed solar cells [3, 4] as a standard AR coating. PS-coated TiO_2 was found to reduce the trapping of surface charge carriers and to enhance the PS photoluminescence stability and the optoelectronic properties of PS-based Si solar cells [5]. Some previous investigations recognized TiO_2 as a promising photocatalyst material for total destruction of common organic pollutants [6]. However, the effective photoexcitation of TiO_2 requires hence irradiation in the ultraviolet (UV) region due to its relatively large band gap (3.2 eV), which leads to a merely 5 % of solar energy absorption. Considerable efforts have been devoted to improve TiO_2 photocatalytic performance in the visible light range. Such efforts include nitrogen, phosphate, chromium and transition metal ions doping, and surface modification with dyes or quantum dots. Nevertheless, challenges payed to reinforce the absorption of the modified TiO_2 materials in the visible range. PS has a large absorption spectrum lying from UV to near infrared, while TiO_2 has absorption limited to an energy radiation as higher as 3.2 eV (corresponding to anatase phase). The adjunction of TiO_2 to PS could in principle generate further excitons that may enhance the photocatalytic activity, by injecting electrons (holes) in the conduction (valence) band of TiO_2, which in turn enhance the photodegradation via an increase of the carrier lifetime.

In the present work, principal experiments concern the discoloration of the Black Amido (BA) during its exposure to UV and visible irradiations when we tested Cr-doped TiO_2 thin films deposited on quarts as well as on multicrystalline porous silicon (PS) substrates.

5.2 Part A: Photocatalytic Activity of Cr-Doped TiO₂ Thin Films Deposited on PS Substrates

To address possible ways to control the photoactivity of Cr-doped TiO_2 thin films deposited PS substrates as well as the structural performance, XRD analysis of such deposited films has been carried out to reach their microstructural features in terms of Cr content.

5.2.1 Microstructural Analysis of TiO₂–Cr/PS/c-Si

5.2.1.1 X-ray Diffraction

XRD patterns of undoped and Cr-doped TiO_2 films, grown on intrinsic silicon substrates at 550 °C, are almost semi-crystalline are displayed in Fig. 5.1. It is found that an anatase-to-rutile phase transition occurs beyond a Cr doping concentration of 2 at.%.

The study of the surface morphology by AFM, the PL properties as well as the effect of PS and TiO_2/PS on the minority carrier diffusion length by means of Light Beam Induced Current (LBIC) measurements have already achieved in Chap. 4.

Fig. 5.1 XRD pattern of
undoped and Cr-doped TiO_2

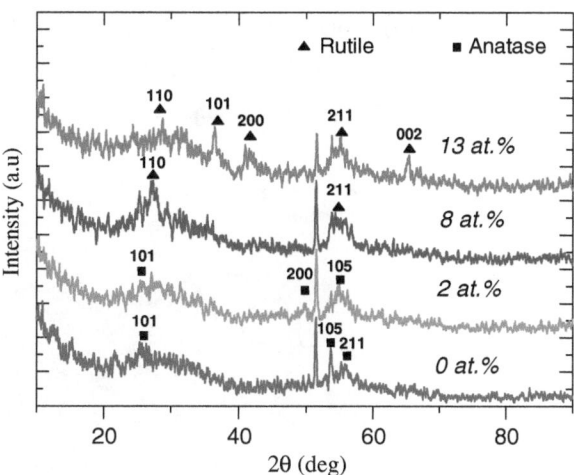

5.2.1.2 Photocatalytic Application

The discoloration of the Black Amido (BA) during its exposure to UV irradiation subsequent to photocatalysis in presence of TiO_2:Cr/PS structure is shown in Fig. 5.2a. Figure 5.2b depicts rather the discoloration kinetics of BA at different Cr concentration. One may notice that the best photocatalytic degradation was obtained for a TiO_2 film with Cr doping of 2 at.% (Fig. 5.2b). It is found that a degradation of the photocatalytic response for Cr content higher than 2 at.%. Indeed, beyond these Cr concentration domain an anatase-to-rutile phase transition occurs (Fig. 5.1); taking into account that anatase is the main active photocatalytic phase in TiO_2, the gradual weakness of the photocatalytic activity of TiO_2 (Fig. 5.2b) may be due to the progressive disappearance of the anatase phase as Cr content increases. It is reported that the photocatalytic degradation of the BA dye can be described by a first order kinetic model [7], $\ln(Co/C) = kt$, where Co is the starting concentration and C is the concentration at time t, k is taken as a constant.

n the same line, the inset of Fig. 5.2b shows the linear transform $\ln(Co/C)$ versus UV light irradiation time exposure. The semi-logarithmic plot of the fitted line was calculated to be $R^2 = 0.99434$ for Cr (2 at.%) doped TiO_2. Figure 5.2c shows the variation of rate constants K versus Cr concentration; K is equal to 0.0076 min^{-1} ± 2.165 10^{-4} for Cr (2 at.%) doped TiO_2. The photocatalytic activity of Cr(2 at.%)-doped TiO_2 is found to be higher than that of the undoped TiO_2 film. This can be related to the surface morphology and to the optical and optoelectronic properties of TiO_2 doped with 2 at.% Cr. In fact, higher surface roughness (RMS) leads to an increase of the effective surface area and may enhance the photocatalytic activity; the RMS of the TiO_2:Cr/PS structures increases from 17 to 43 nm as Cr doping varies from 0 to 4 at.% as described before (Chap. 4). Good photocatalytic response at an optimal concentration of 2 at.% Cr is consistent with the increase in the minority carrier lifetime probably coming from carrier exchange with PS.

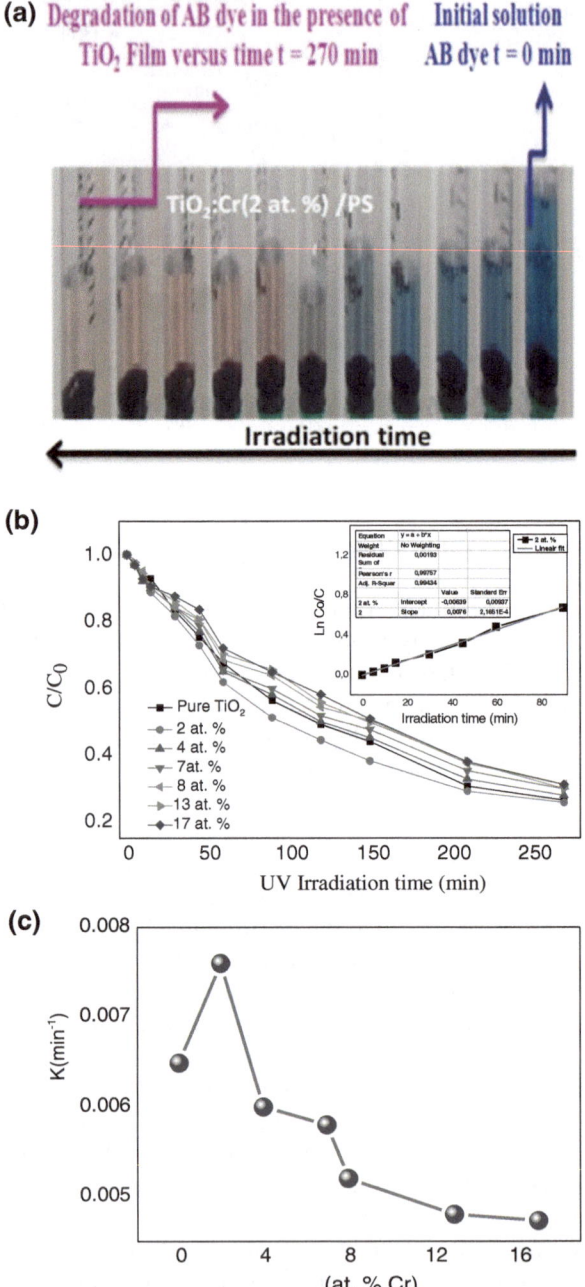

Fig. 5.2 **a** Black Amido (*BA*) discoloration with TiO$_2$:Cr/PS, **b** evolution of the BA degradation as a function of UV light irradiation time for TiO$_2$:Cr/PS films having different Cr concentrations; the *inset* indicates the linear transform ln(Co/C) versus UV light irradiation time for Cr (2 at.%) doped TiO$_2$/PS, **c** kinetic constant related to the discoloration of BA versus Cr concentration

5.3 Part B: Photocatalytic Activity of Cr-Doped TiO$_2$ Thin Films Deposited on Quartz Substrates

5.3.1 Optical Analysis TiO$_2$–Cr/Quartz Substrate

5.3.1.1 Urbach Energy and Absorption Coefficient

It is found that the gap energy varies between 3.3 and 2 eV with Cr content increases [8], with an abrupt transition around 7 %. This variation may be related to the presence of acceptor states in the bandgap. Transition metals could also make significant changes on the electronic structure of a crystalline material and thus on the value of the gap [9, 10]. The decrease in the gap makes Cr-doped TiO$_2$ thin films more absorbent on a wide range of UV-Visible, constituting a major advantage in terms of other applications such as photovoltaic ones.

Dopants generally provide Schottky barrier which facilitates the transfer/trapping of electrons from TiO$_2$ and improves the photocatalytic efficiency. To prove this effect the modification of the band structure of a TiO$_2$ due to introduction of Cr may be reached by determining the Urbach energy (Eu) before and after introduction of the dopant. Urbach energy provides a measure of possibly structural disorder in a material. The formula used to determine the Urbach energy is given by [11]:

$$(\alpha h\nu) = \alpha_0 \exp\left(\frac{h\nu}{E_U}\right) \tag{5.1}$$

The extinction coefficients (Fig. 5.3) as estimated by previous ellipsometry characterization (Chap. 4) are used to calculate the absorption coefficient (α) versus photon energy ($h\nu$) by the following formula [12]:

$$\alpha = \frac{4\pi K}{\lambda} \tag{5.2}$$

Fig. 5.3 Spectral dependence of α for TiO$_2$ films as a function of Cr concentration

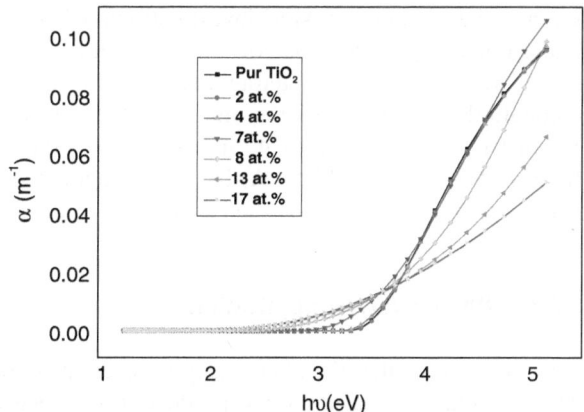

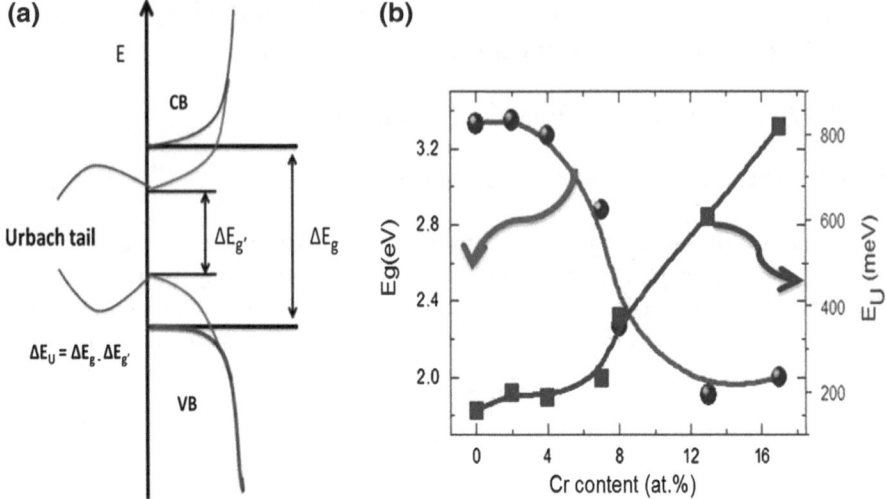

Fig. 5.4 Cr doping effect of on the band gap evolution: **a** Doping induced smearing of the valence and conduction band edges and formation of the Urbach tail, **b** variation of the band gap and Urbach energy for different Cr concentrations

where k is extinction coefficient and λ is the wave length of an incident photon energy.

The Urbach energy values were calculated for undoped and Cr-doped TiO₂ thin films (Fig. 5.4).

The Urbach energy corresponding to undoped TiO₂ film is about 0.156 eV, while the Cr-doped film has 0.818 eV as Urbach energy especially for film containing 17 at.% of Cr element. This is caused by the presence of impurity levels introduced by Cr doping in the band structure of undoped films. Since higher Urbach energy is indicative of considerable introduction of tail states at the band edges, the higher E_u value of Cr-doped films indicates further introduction of tail states as compared to undoped ones. Moreover, it is found that a critical Cr doping around 4 % gives a minimum E_u value which in turn corresponds to the maximum grain size as detailed in AFM investigations (Chap. 4). This may be due to the formation of CrO₂ oxide phase (a = 4.21 Å and c = 2.91 Å) that crystallizes in the same structure than TiO₂ (tetragonal structure), therefore leading to a high lattice compatibility between doping and host matrices as previously reported [13, 14]. When Cr content increases it, appears rather other related oxides such as Cr₂O₃ (a = 4.95 Å and c = 13.58 Å) having hexagonal structure, (so different to TiO₂) leading to high E_u value and then a decrease of the grain size.

5.3.1.2 Photocatalytic Application

It is well known that the photocatalytic reaction is sensitive to the catalyst surface. The reaction is initiated by the production of electron-hole pairs on TiO₂ surface resulting from the absorption of photons with energy equal or greater than the band

Fig. 5.5 Spectra of the absorbance of Amido Black (10 mg/l) at various degradation time using Cr (2 at.%) doped TiO₂ film

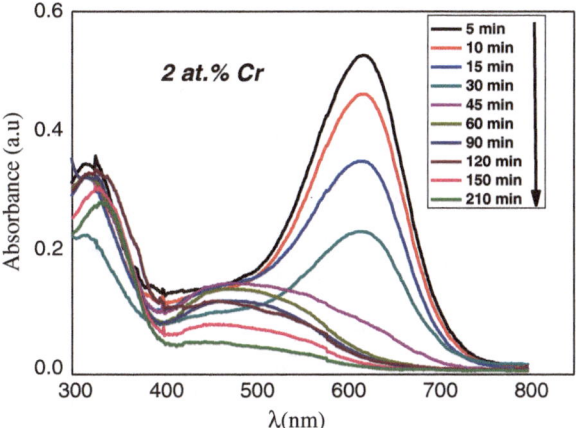

Fig. 5.6 Evolution of the Amido Black as a function of UV light irradiation time for Cr-doped TiO₂ films having different Cr concentrations. *Inset* linear transform ln(Co/C) versus UV light irradiation time for Cr (2 at.%) doped TiO₂

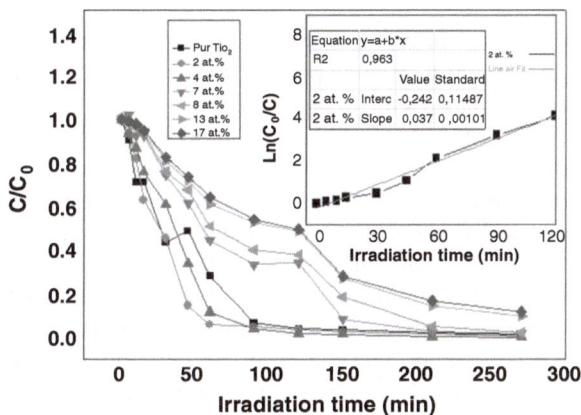

gap energy. Amido Black (AB) reacts with the electrons generated on the TiO₂ particles under UV irradiation. Figure 5.5 show the absorbance of AB at 618 nm as a function of UV light irradiation time for TiO₂ films doped with Cr 2 at.%. To demonstrate the degradation process, the spectra were recorded at different interval times. The experiments were repeated with two sets of films annealed at similar conditions and the results are found to be reproducible. Figure 5.6 shows the kinetic measurement of the effect of Cr content on the C/Co under light irradiation, indicating the photo-degradation of AB and inset Fig. 5.6 show the Linear transform of ln (Co/C) as a function of UV light irradiation time for Cr (2 at.%) doped TiO₂.

It is pointed out that the photocatalytic degradation of AB dye can be described by a first order kinetic model, **ln Co/C = kt** [7], where Co is the initial concentration and C is the concentration at time t and k is the kinetic constant.

Figure 5.7 shows the increase of the photocatalytic activity for Cr concentration varying in the 2–4 at.%. This can be related to the surface morphology and to the crystal structure. In fact, RMS of TiO₂: Cr films increases from 1.5 to 3 nm as the

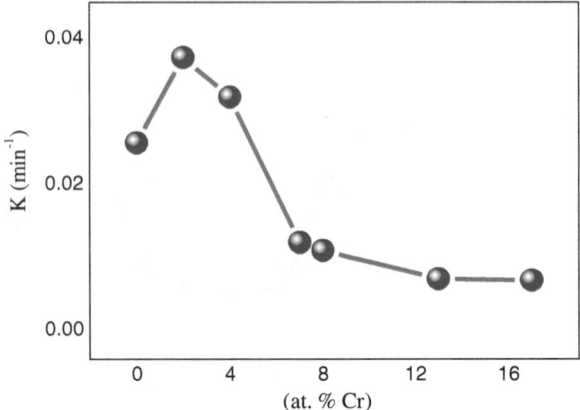

Fig. 5.7 Kinetic constant related to the discoloration of Amido Black versus Cr concentration

Cr content varies from $0°$ at.% to $4°$ at.% (related to anatase phase). Higher surface roughness leads in principle to an increase in the effective surface area, which in turn enhances the photocatalytic activity.

It is noted that the photocatalytic activity rate reaches an optimum for sample doped with 2 at.% of Cr. This optimum could be explained by the fact that the presence of dopant prevents the recombination of exciton. Further increase in Cr content decreased the % degradation of AB. This is due to the increased coverage of the semiconductor surface by the metal which decreases the exposed surface to the dye. The diminished penetration depth of light at higher concentration could either prevent the direct excitation of semiconductor.

It is assumed that the observed optimum concentration of Cr is assigned to a combination of an optimum distribution of Cr nanoparticles (density) on the TiO₂ surface and to an optimum Cr nanoparticles size. The effect of size on photoactivity well known phenomena. Matko et al. [15] and Henry [16] showed that very fine Pd/Pt particles are more active than the big ones. This size effect can be understood in terms of the presence of additional adsorption sites for the dye at the particle surface. The increase of the number of surface atoms which have dangling bonds can serve as a site for adsorption. Moreover, it is reported that oxidized small metallic nanoparticles are more active than reduced one. The level of oxidation state increases with particle size.

For big Cr particle size the catalytic effect is only superficial and will not affect noticeably TiO₂ electronic structure. Very high dispersion of dopant nanoparticles on TiO₂ surface will also lead to small effect, (Fig. 5.8).

It clearly appears that a probably much better sensitivity would be achieved if the aggregate distribution in the film could be specifically characterized by a very small particle size coupled with a high density.

The vast oxygen vacancy formation occurring with Cr doping could also contribute to the photocatalytic activity enhancement of TiO₂. Cr-doping introduces

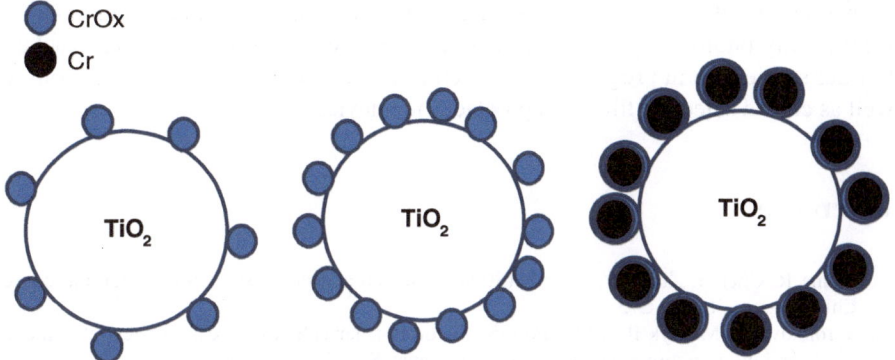

Fig. 5.8 Schematic distribution of Cr doping nanoparticles on TiO$_2$ surface when Cr content increases

Fig. 5.9 Effect of P$_{Cr}$ on the ratio of oxygen concentration related to TiO$_2$ and oxygen concentration related to CrOx

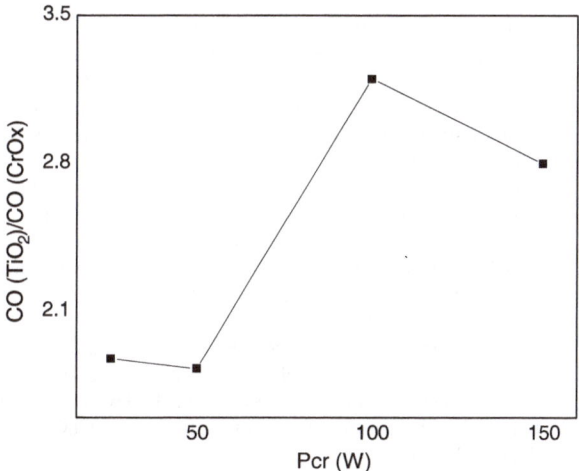

substantive oxygen vacancy as reported by XPS analysis (Fig. 5.9), the oxygen vacancies in TiO$_2$ act as electron traps which can bind the photoinduced electrons and play a significant role in inhibiting the recombination rate of photoinduced electron–hole pairs, thus enhancing the photocatalytic activity of TiO$_2$ [17].

5.4 Conclusion

Cr doped TiO$_2$ thin films have been grown on quartz and PS substrates. These films are characterized especially by means of various structural, optical as well as LBIC techniques. It is found that such investigations emerged as a new way to show the structural modification of thin films microstructure and show potential properties in photocatalytic tests. These results are very interesting since a significantly photocatalytic activity based on simple structures have been reached. Further studies are in progress to use such films in other sensitivity applications as well as coated selective films in optoelectronic devices.

References

1. Zhang R, Chen Y, Zheng Y, Chen L (2009) Research and progress of Silicon luminescence. Chin J Lasers 36:269–275
2. Yerokhov VY, Melnyk II (1999) Porous Silicon in solar cell structures: a review of achievements and modern directions of further use. Renew Sustain Energy Rev 3:291
3. Martinu L, Poitras D (2000) Plasma deposition of optical films and coatings: a review. J Vac Sci Technol, A 18:2619–2645
4. Chatterjee S (2008) Titania–Germanium nanocomposite as a photovoltaic material. Sol Energy 82:95–99
5. Kim KJ, Kim GS, Hong JS, Kang TS, Kim D (1998) Characterization of a composite film prepared by deposition of TiO$_2$ on porous Si. Sol Energy 64:61–66
6. Bahnemann D (2004) Photocatalytic water treatment: solar energy applications. Sol Energy 77:445–459
7. Wenhua L, Hong L, Saoan C, Jianqing Z, Chunan C (2000) Kinetics of photocatalytic degradation of aniline in water over TiO$_2$ supported on porous nickel. Photochem J Photobiol A: Chem 131:125
8. Hajjaji A, Gaidi M, Bessais B, El Khakani MA (2011) Effect of Cr incorporation on the structural and optoelectronic properties of TiO$_2$:Cr deposited by means of a magnetron co-sputtering process. J Appl Surf Sci 257:10351
9. Luu CL, Nguyen QT, Ho ST (2010) Synthesis and characterization of Fe-doped TiO$_2$ photocatalyst by the sol-gel method. Advances in natural sciences. Adv Nat Sci: Nanosci Nanotechnol 1:015008
10. Singh D, Singh N, Sharma SD, Kant C, Sharma CP, Pandey RR, Saini KK (2011) Bandgap modification of TiO$_2$ sol-gel films by Fe and Ni doping. J Sol-Gel Sci Technol 58:269
11. Arushanov E, Levcenko S, Syrbu N, Tezlevan V, Merino M, Leon M (2006) Urbach's tail in the absorption spectra of CuIn$_5$S$_8$ and CuGa$_3$Se$_5$ single crystals. Phys Sta Sol (a) 203(2011):2909
12. Sharma P, Katyal SC (2006) Influence of replacing Se in Ge10Se90 glassy alloy by 50 at. % Te on the optical parameters. J Ovonic Res 2:105
13. Boubaker K, Amlouk M, Louartassi Y, Labiadh H (2013) About unexpected crystallization behaviors of some ternary oxide and sulfide ceramics within lattice compatibility theory LCT framework. J Aust Ceram Soc 49:115
14. Boubaker K, Amlouk M (2013) Amorphous ternary ceramics instability below 450°C: nano-scale arguments from the lattice compatibility theory (LCT). Int J Appl Ceramic Tech 11:1
15. Matko I, Gaidi M, Hazemann JL, Chenevier B, Labeau M (1999) Electrical properties under polluting gas (CO) of Pt- and Pd-doped polycrystalline SnO$_2$ thin films: analysis of the metal aggregate size effect. Sens Actuators B 59:210
16. Henry CR (1989) On the effect of the diffusion of carbon monoxide on the substrate during CO oxidation on supported palladium clusters. Surf Sci 223:519
17. Peng Y-H, Huang G-F, Huang W-Q (2012) Visible-light absorption and photocatalytic activity of Cr-doped TiO$_2$ nanocrystal films. Adv Powder Technol 23:8–12

Conclusion

Current Status and Perspectives for Chrome-Doped TiO_2 Thin Films

During this work we were interested in the development, characterization and application of titanium dioxide thin films doped with Cr (TiO_2–Cr) obtained by sputtering. A detailed study was made on the effect of heat treatment and the Cr incorporation on the microstructural and optoelectronic properties of the realized layers. A parametric study was undertaken in order to get a finely control of the different properties for the developed films. Thus the influence of the deposition parameters (Cr concentration, effect of deposition temperature, development time, power) were closely correlated to the microstructural, optical and electrical characteristics.

Furthermore, the analysis of the microstructural and optical properties of the Cr doped and non-doped TiO_2 thin films, shown that titanium dioxide films crystallize in two phases, rutile and anatase, at an annealing temperature of 550 °C. The anatase-rutile phase change takes place for a Cr doping of 7 % and an annealing temperature of 550 °C; the rutile phase (more stable) is achieved at temperatures over 700 °C.

The elaborated layers have been incorporated in a gas sensors structure and a photovoltaic structure. We have shown that the sensors are fairly stable, and TiO_2 film doped with Cr (4 %) presents the best response to ethanol. Furthermore, at 0.1 % ethanol, a maximum sensitivity is reached for an operating temperature of 200 °C. It has been also shown that titanium dioxide obtained by sputtering and deposited on porous silicon substrates formed initially from monocrystalline and multicrystalline substrates can contribute effectively to the reduction of reflection losses. The deposition of a Cr doped TiO_2 film on PS enhances the diffusion length and lifetime of the silicon substrate.

© The Author(s) 2015
A. Hajjaji et al., *Chromium Doped TiO₂ Sputtered Thin Films*,
SpringerBriefs in Manufacturing and Surface Engineering,
DOI 10.1007/978-3-319-13353-9

In conclusion, this study shows that Cr-doped TiO_2 films combined with porous Silicon is a treatment capable of improving mono and multicrystalline Si optoelectronic properties. A dramatic increase in the lifetime of the multicrystalline Si equal to 733 μs was reached. The experimental results suggest that the TiO_2–Cr/PS induces a good multicrystalline and monocrystalline silicon surface passivation.

As perspective for this work, we suggest the deposition of ohmic contacts on different layers of the Cr-doped titanium dioxide and study of the optoelectronic properties such as quantum efficiency and the photo-courant. Other possibilities would be to improve the electrical properties of these layers by the introduction of other dopants (Pt, Fe,…) and check their use in pollution sensors.